Keeping Ducks, Geese and Turkeys

Garden Farming Series

Keeping Ducks, Geese and Turkeys

John Walters and
Michael Parker

PELHAM BOOKS

First published in Great Britain by
Pelham Books Ltd
27 Wrights Lane
London W8 5TZ
1976
Reprinted 1978
Paperback edition 1982
Reprinted 1985
Laminated hard cover edition 1987

British Library Cataloguing in Publication Data

Walters, John, 1944–
    Keeping ducks, geese and turkeys.—
    3rd ed.—(Garden farming series).
    1. Poultry—Amateurs' manuals
    I. Title II. Parker, Michael, 1933–
    III. Series
    636.5'083   SF487

ISBN 0–7207–1791–4

Printed in Great Britain by
Hollen Street Press Ltd, Slough

To
Nick and Lucy, Sarah and Andrew

Contents

Acknowledgements

There are only two names on the front cover of this book, but a number of others have had a hand in its production. We would like to thank most particularly Derek Kelly of Danbury, Essex, for his advice on turkeys; John Woods of Scorton, Lancs., for invaluable advice on the rearing of geese; and Fokko and Peter Kortlang of Ashford for their advice on ducks.

Our thanks also to the British Waterfowl Association and their secretary Christopher Harrisson for help in compiling a list of suppliers of goslings and ducklings and to the breeders who allowed us to include their names.

Finally, we would like to thank our families for their tolerance while the book was being written and our respective wives for their own contributions—to Jane Walters who, with assistance from Colin Reeves, turned our very amateurish sketches into finished line drawings, and to Jan Parker who had to decipher our handwriting to produce the final typescript.

Conversion Factors

Throughout the text Imperial units (e.g. feet, inches, pounds, ounces, °F) have been used for measurements of houses, runs, feeds, temperatures, etc., since most of us are still a long way from 'thinking metric'. For information a useful table of conversion factors is given below.

Imperial Units to Metric Units

1 in.	=	2.54 cm
1 ft	=	0.3048 m
1 mile	=	1.6093 km
1 gal	=	4.5459 litres
1 sq. in.	=	6.4516 sq. cm
1 sq. ft	=	0.0929 sq. m
1 sq. yd	=	0.836 sq. m
1 acre	=	0.4046 hectare
1 oz	=	28.34 g
1 lb	=	0.4536 kg
1 ton	=	1.016 tonnes
$x°$ Fahrenheit	=	$\frac{5}{9}(x-32)°$ Centigrade

Metric Units to Imperial Units

1 mm	=	0.0394 in.
1 cm	=	0.3937 in.
1 m	=	3.2808 ft
1 km	=	0.6214 miles
1 litre	=	0.2199 gal
1 sq. cm	=	0.1550 sq. in.
1 sq. m	=	10.764 sq. ft
1 sq. m	=	1.195 sq. yd
1 hectare	=	2.471 acres
1 g	=	0.035 oz
1 kg	=	2.2046 lb
1 tonne	=	0.9842 tons
$y°$ Centigrade	=	$(\frac{9}{5}y+32)°$ Fahrenheit

Grow Your Own Sunday Lunch

General considerations

Ducks and geese can be attractive, endearing creatures. Turkeys, with their more austere, forbidding features, are less likely to gain a ready place in one's affections. But however firmly entrenched any one of the three species becomes as part of the family, it should be remembered that it is being kept for two reasons only: to save money by making the housekeeping stretch that bit further; and to give a semblance of self-sufficiency in an age when it is necessary to depend on so many outside sources for our daily requirements. Apart from all that, there is the undeniable satisfaction of entertaining friends and family to a Sunday lunch, or even to the traditional feast at Christmas, that you have 'grown' yourself. Many families can boast that the vegetables came out of the garden, but how many can add that the meat—be it duck, goose or turkey—did as well?

'Grow your own Christmas dinner' can be a splendid slogan with which to start your garden poultry farm, but it must not cloud the fact that there is a lot of work attached. It is not so much hard work, but an unremitting chore that has to be done seven days a week, regardless of the weather. If, for any reason of holiday or sickness, you are unable to look after your birds, you have to make sure there are friendly neighbours willing to step into the breach.

Having established that profit is the motive behind the enterprise, it is essential that you accept the fact that the birds will only be part of the family for a relatively short time. If it is difficult to reconcile the family to the fact that the birds have to be killed eventually, it might help to remind them that the birds would not be there at all if it were not for their ability to turn grain into meat or eggs so efficiently.

The choice as to which type of stock will depend very much on the size of your garden and on your own taste, i.e. whether you prefer duck, goose or turkey on the table, or duck eggs to chicken eggs.

It is impossible to be dogmatic on which is the best type of meat—that again will depend on individual preference. If you prefer taste and flavour, ducks or geese will probably be the choice, but for a given weight they produce less meat. A 4 lb oven-ready duck, for instance, will feed three people at most; four hungry people will need a couple of 4 lb ducks to satisfy their appetites.

Ducks

With ducks, of course, you have two choices: meat or eggs. This again will depend on taste. Duck-egg eating has never caught on to the same extent as chicken-egg eating, which is a pity, for the duck is a magnificent egg machine. Egg-laying breeds, like the Khaki Campbell, regularly clock up 300 eggs a bird a year. The average yield of about 230 eggs a bird a year is on a par with the intensively housed and highly developed chicken. Duck eggs tend to be larger than chicken eggs and therefore give a greater total egg weight; on the other hand, ducks eat more!

Duck eggs have a strong flavour which may not be to everyone's taste. They are also burdened by the slur of salmonella following a number of food-poisoning outbreaks. The basic cause of the salmonella scare is the shell of the egg. It has relatively large pores, picks up disease organisms easily and gives the egg relatively poor keeping qualities. A chicken egg stored at around 50°F will keep fresh for about three weeks; a duck egg should be eaten within a week, or ten days when kept at the right temperature.

Another reason why duck eggs are seldom seen for sale could, of course, be their complete lack of sales promotion. As a marketing expert once told a sales conference in Warwickshire, 'Only the duck, when she has laid it, makes a noise about her egg—and who buys duck eggs?' The answer, unfortunately, is very few. Annual consumption is about 3·8 million dozen compared with 1,130 million dozen chicken eggs.

Geese

Much of the flavour in geese comes from their generous allowance of fat, but this can be a drawback both when the goose reaches the table and when it encounters really hot weather in summer. Wild geese fly north in the summer to breed in the Arctic, but the domestic type has no such opportunity—it has to stay south and sweat it out.

At one time, a goose was the only bird to have at Christmas, but lack of attention to quality and too many birds with too much fat toppled the goose from its No. 1 spot. Today, over ten million turkeys are eaten at Christmas, while only 200,000 geese are eaten throughout the year.

The fat problem can be overcome by choosing the right stock and by not running the geese on too long. Killing at the optimum weight and age will prevent a build-up of fat and uneconomic use of feed. The fat can, however, be turned into a virtue. It is said to be ideal for all sorts of culinary uses: better than butter in baking; a provider of flavour to vegetables, fried meats, gravies and sauces; and without a peer for deep-frying chipped potatoes.

As a rough guide, 10 per cent of the total weight of a goose is lost in blood and feathers and 20 per cent in intestines, head and feet, so that a 10 lb live-weight bird will be 7 lb when ready for the oven. (Appendix

E, indicates how weight is lost with each process.)

Turkeys

What turkeys lack in flavour they make up for in economy. While it takes something like 4 lb of feed to produce 1 lb of duck or goose, the turkey will put on 1 lb in weight for every $2\frac{1}{2}$ lb of feed, give or take an ounce or two. This is just one of the benefits accruing from the years of research that have gone into turkeys. In a little over four months, a turkey stag will reach 16–17 lb live-weight.

Turkeys do not make as much noise as ducks or geese, but the stags will gobble, so go for the quieter hens if your unit is placed near neighbours who might otherwise object.

Garden space requirements

If you have a garden of less than $\frac{1}{4}$ acre, geese are out. They are only an economic proposition if they can graze and even with $\frac{1}{4}$ acre of suitable grass, there will only be enough grass for about twelve geese. Without that much grass, a compounded feed will have to be given and the poor conversion ability of the goose makes the practice a non-starter.

Non-grazing ducks and turkeys can get by on less space and 100 sq. ft of lawn will take six of either species. The problem of how to avoid turning the lawn into a mud patch is covered in the housing chapter (pages 38–49).

Economics

Finally, what will it cost to set up your flock in business? The prices given in the shopping list (Appendix B, page 116) were those operating at early 1985 levels and can be a guide only.

Stock can be bought at a number of ages: at day-old; at week-old; or at four weeks of age when they are off heat. Naturally, the price increases with the age of the bird, but off-heat stock is an investment worth considering for the novice.

Turkey breeders will sell poults (young turkeys) at a slightly cheaper rate outside their busy months of July and August. The price of poults will also vary according to the quantity bought.

In Appendix B the prices given for the three types of stock include carriage, but those for feed are on a cash and carry basis. Delivery could put an extra £1.00 on the bill so it obviously pays to collect the feed yourself if possible.

Local millers and agricultural feed merchants are the best sources of supply of feed and generally give the best prices. Poultry farmers' supply companies are also useful and if there is one locally he should be listed in the *Yellow Pages*. Obviously, it is better to buy in as large quantities as possible, but this will depend on what storage space you have available.

Many millers and merchants do not stock turkey rations in the first half of the year, i.e. before the farmers start growing their flocks for the Christmas trade, but a special order will be fulfilled.

The most expensive places to buy feed are the high-street pet-shops. They may not always stock turkey rations but you can usually place an advance order with them.

At this stage it must seem that a lot of money is going out and the prospect of any coming in by way of a return on the investment must seem remote. But returns, of course, will mainly be evident at meal times and their cash value can only be gauged by working out the current retail price of the bird you are eating.

These prices will vary according to whether the birds are fresh, frozen, oven-ready or traditional farm-fresh plucked, but an allowance of 90p a lb oven-ready will give a rough guide to the value of your 'home-grown' meal. If you sell off surplus birds to friends and neighbours, keep a close check on prices in your area so you have an idea of what to charge, not forgetting that a fresh bird will always command a premium over the frozen product.

Restrict your marketing to friends and neighbours, and steer clear of butchers or other retail outlets unless you want to get involved in extra administration and cost. Before selling to anyone outside your immediate circle of friends, contact your local Ministry of Agriculture Advisory Officer (see Appendix A) for details of the latest marketing regulations.

The market return on duck eggs is in the region of 85p a dozen, but this again will vary according to the time of year, so check local levels if you are selling any.

Geese eggs can also yield a return, of course, and at a retail price of about 30p each this is not to be sneezed at. Although too large for the average egg-cup, they make the lightest, fluffiest omelettes and are ideal to use in cooking. Average size is 7 oz—the weight of three Size 2 chicken's eggs.

Two other by-products from geese are the feathers and the down. The down can be used for stuffing quilts, pillows and sleeping bags and the feathers for quill pens (such a pity they have been superseded by the ball-point). A top-weight Embden or Toulouse goose will yield up to $\frac{1}{2}$ lb of feathers.

Two final points before you order the first birds and start building a house and pen—make sure that there are no local regulations which forbid the keeping of farm livestock, and that your immediate neighbours have no overwhelming objections to the enterprise. It is important to keep on the right side of the neighbours if you have plans to engage their help at holiday times, so it would seem to be both diplomatic and polite to secure their initial blessing. The local council offices will put you right on the regulations.

Now, assuming you have surmounted those two hurdles, on with the business of setting up your garden duck/geese/turkey farm. Have fun!

Management Of Your Flock

Ducks

Ducks have four basic needs if they are to flourish and give a good return in terms of eggs or meat. They are:

(1) Heat in the vital first three weeks of life.

(2) A draught-free, dry house that will protect them from bad weather and predators.

(3) A balanced diet that caters for all their nutritional needs.

(4) Enough feed and enough trough space so that if all the flock decides to eat or drink at the same time there is enough room for the smallest and the most timid member.

Follow these four points and the flock will reach either killing age or point of lay without any checks on growth rate and you will have the same number of ducklings alive at the end as you started with. Inevitably some birds will thrive less well than others, and others will die accidentally, but good careful management will keep the number of deaths to a minimum.

The best stockmen add an ingredient which goes under the initials T.L.C. They letters stand for 'tender, loving care', and it is amazing how well a flock will flourish and respond to a little affection. But do not confuse affection with coddling; young stock will survive the most bitter conditions so long as their basic needs are met.

Whether you keep ducks for the eggs they lay or the meat they produce, rearing in the initial three to four weeks will be more or less identical. An egg-laying Khaki Campbell or a meat-producing Aylesbury will both need heat when they are day-old ducklings. This can either come from the natural source—a broody duck—or by artificial means—electricity, gas or paraffin. To keep it simple for the back garden, we will concentrate on the use of a broody duck and the infrared lamp.

Natural brooding

A broody is ideal for small numbers. She will take up to fifteen ducklings, or about ten eggs if she is hatching them as well. Mum and her brood will need separate accommodation from the main flock, so it will be necessary to provide a brooder house for them. If, however, these ducklings are to form the basis of your first flock then, of course, the main quarters can be used.

A 6 ft × 4 ft shed, if available, would be just the place in which a dozen ducklings could spend the first four weeks of their lives. Give the floor a covering of peat moss, sawdust, shavings or short cut straw (which should be renewed daily) and allow the ducklings access to a small range area—about 48 sq. ft. At three or four days old, the stock will be ready to move outside if the weather is reasonable, but don't let a biting east wind be their first experience of the great outdoors.

A 2 ft high wire-netting surround will prevent them wandering too far, and if there are crows about, a netting roof will be necessary. One of the less endearing traits of crows and magpies is to 'dive-bomb' ducklings and nip their heads off, a vicious habit definitely to be discouraged.

Initially, food troughs and drinkers should be kept

inside the brooder house, but when the youngsters start to venture forth, their food supplies should go with them.

In the early days, it is unlikely that you will run a breeding flock, so the natural brooding of ducklings will have to be done by a recruit from an adult flock. The heavier strains make the best mothers and they have an advantage in that their bulk will accommodate more ducklings. 'Mother' Aylesbury at around 9 lb will be able to keep more youngsters warm than 'Mother' Campbell at about 5 lb. If a broody hen is available from a local flock of chickens she can be used—in fact, a hen can be more successful; ducks with their webbed feet can make clumsy mothers and ducklings do get crushed from time to time.

Ducks are good-natured creatures, but they should not be taken for granted. If you are going to use a 'foster' broody, take care over the initial introduction: first impressions are as important in the duck house as anywhere else.

Indeed, introductions should start before the ducklings arrive. Present your waiting broody with a clutch of eggs to sit on for a week beforehand—china eggs can be bought for this very purpose. Pre-arrival preparations should also include a delousing of the 'mother' with a dusting powder before installing her in a nest-box, if one is available.

Evening is the best time for introductions, so unless the ducklings are very hungry and fighting to get out, leave them in the box in which they arrived. But don't leave it so late that you find yourself installing the last ducklings after midnight. The complete process can take hours.

Put just one duckling under the foster mother initially, close the front of the nest-box and leave them together for about forty-five minutes. As a measure of the duckling's reception, scatter some corn before the hen and watch to see whether the youngster is invited to join the feast.

If all is well and the invitation to dine is forthcoming, put in two or three more ducklings and repeat the process every thirty minutes or so until all the new arrivals are installed. If, on the other hand, the first duckling fails to be accepted, give 'mum' a thirty-minute break and repeat the process.

The mother duck should stay with her brood for two or three weeks, after which she can return to the adult flock. The time will depend on the season and the outside temperature; in summer, brooding time will be near two weeks, but probably three weeks in winter.

The infra-red system

Obviously, the house will need electricity laid on for this method of brooding, which will be an added expense. But once the power is there, it can be invaluable. As always with electricity, its installation should be left to experts.

The infra-red system is ideal for small numbers of birds and one 250-watt lamp will provide warmth for up to thirty or forty ducklings. If you exceed this number, you will need two lamps set 18 in. apart above the ducklings. Day-old birds need a temperature of 85–90°F, which can be obtained by adjusting the height of the lamp above the floor.

From the initial temperature of 85–90°F, the brooder heat should be reduced by five degrees every three days; this is achieved by raising the height of the lamp. In summer the lamp can be switched off after about ten days, but it will be necessary to keep the brooder on longer in the winter. How much longer will depend on the weather, but no more than four or five days extra unless the weather is bitterly cold.

The day before the ducklings are due, set a spirit thermometer on the litter floor 6 in. from the centre of the lamp, then adjust the lamp's height until you get a correct reading. Spirit thermometers are more reliable than the mercury type for this job.

A big advantage with the infra-red system is that the ducklings remain in view all the time; there are no curtains, as there are with some systems, to obscure the view. The birds soon give an indication if the lamp height is wrong. If they pant and appear distressed, the lamp is too low. If they huddle in groups, instead of spreading out evenly at night, it is too cool. In both cases the temperature should be checked immediately. The 'hay-box' system for goslings, described on page 33, can also be used for a small flock of ducks.

Another point to check is the reflector. If a dull emitter type of lamp is used and the reflector becomes dusty, its effectiveness will be impaired, so take care to keep it bright and clean. With the bright emitter type, the reflector is built into the lamp where it cannot get dirty.

For the first two or three days, an 18 in. high corrugated cardboard surround will prevent the ducklings wandering too far from the source of heat (see Fig. 1), but remember to extend the area after three or four days. The young stock will grow rapidly and a huddle of them against the surround is a sure sign that they are in need of extra space. Normally they should have the run of a 6 ft × 4 ft house by the beginning of the second week, though again it will depend on the time of year.

Always be generous with space, never confine ducklings too closely. If they start to strip the down from each others' backs, there are too many. The minimum space requirements in the first ten weeks are shown overleaf.

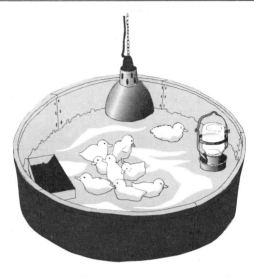

Fig. 1 Housed with a 250-watt infra-red lamp with reflector, a jam-jar drinker and feed in a shallow tray, ducklings will thrive. The cardboard surround should be moved back as they grow and the feeding and drinking points kept beyond the range of the lamp.

Age in weeks	Sq. ft allowance per duckling
0–2	$\frac{1}{2}$
2–4	1
4–6	2
6–8	3
8–10	4

Drinkers and feeders

Feeding of young stock is discussed in Chapter 5, but a few words on the types of feeders and drinkers to use is relevant here. In the first two or three days, the feeders need be no more sophisticated than cardboard lids or egg trays. Tin lids can also be used, but avoid ones with sharp edges which will cut the birds' feet. Also to be avoided are china plates or saucers as feet are liable to slip on the glazed surface.

By the end of the first week, however, they will have outgrown lids and egg trays and will be ready for proper troughs designed to prevent waste. Special troughs can be bought for young stock, but they soon outgrow them so it is more economical to start them with an adult version. There are a number of proprietary feeders (and drinkers) on the market and details of the manufacturers are found in Appendix A at the back of the book.

Wet mash should be fed in a trough with a bar running a couple of inches above the level of the feed. This will prevent the ducklings from fouling the feed with their droppings and also from treading in the mash and scattering it all over the floor. Troughs are easy enough to make and a simple example is shown in Fig. 2. It should be 5 in. wide and $2\frac{1}{2}$ in. deep. Lips running down either side will help prevent waste.

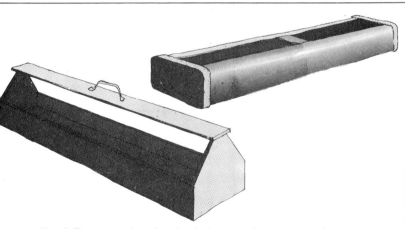

Fig. 2 Two examples of a simple home-made range trough. Width should be about 5 in. and depth about $2\frac{1}{2}$ in., but do not make either type too long or they will be awkward to carry. Two 2 ft troughs will be easier to handle than one 4 ft long.

It is vital to provide enough trough space so that every bird gets a chance to feed. The amount will vary according to the type of feed and the age of duckling. During the first three weeks allow 3 in. of trough space per duckling when feeding wet mash and 2 in. for crumbs. A trough where the ducklings can feed from both sides will double the feeding space, i.e. a 2½ ft trough will have 5 ft of feeding space.

A constant supply of drinking water is another must and there is a wide range of equipment available. In the early stages, avoid drinkers where the water is deep enough for the ducklings to immerse themselves. On the other hand, provide enough water for them to submerge their heads and beaks. Without this facility, heads and eyes become scaly and crusty and the ducklings cannot wash their bills clear of feed. Examples of various types of water fount and automatic waterer are shown in Fig. 3. Grids are necessary over the open trough to keep the birds out of the water.

You will soon learn that the water needs regular replenishing—once a day at least. Ducks and geese have a habit of taking a bill full of feed then waddling over to the water trough to rinse it down. Naturally, some of the feed is washed off into the drinkers and quickly builds up, leaving a murky liquid by the end of the day.

Drinkers and food troughs must not be placed under brooder lamps. Obviously they must be within the cardboard surrounds, but keep them clear of the actual heat. There is a theory that bright emitters in the brooder lamp destroy certain vitamins in the ration, but apart from this, brooder-warmed feed and water can be unpalatable. If the ducklings will not come from under the brooder to feed or drink it shows that the birds are not warm enough and the temperature should be checked (see page 22).

Later, when in a grass pen, feeders and drinkers

should be kept on the move to prevent patches being worn.

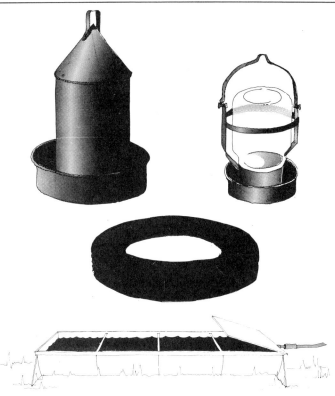

Fig. 3 Four types of drinker. *Top left:* the half gallon version is suitable for stock of all ages. If used with very young stock, pebbles or marbles can be placed in the base of the trough to reduce the depth of the water. *Top right:* the upended jam-jar fitted to a patented trough base is ideal for stock up to about two weeks old. *Centre:* cheapest of all is the old tyre cut in half—a saw, a drop of oil and some elbow grease are all that's required. Each half will cater for about fifteen ducks. *Bottom:* the mains-fed pig drinker has a ball valve to control the level of the water. Wire netting (about 2 in. gauge) will be needed over the top to keep ducklings out of the water.

Daily management

Ducklings are easily frightened, timid creatures and management should be tailored to take care of this fact—no sudden shocks, no nasty surprises. Let everything happen gradually and at the same time every day, as far as possible.

Always give them warning of your approach by whistling, singing or chatting to the birds as you near the house. You may feel a bit daft as you tra-la-la your way down the garden in the pouring rain, but it will pay dividends in terms of even, uninterrupted growth. If the flock runs to one end of the house every time you approach, it is a difficult habit to break and it can go on for weeks. Apart from any approach you may adopt, a gentle knock on the door before entering will also calm their fears.

Put a firm ban on all visitors, particularly young ones, during the rearing period. This will be difficult and is not calculated to win you friends, but unfamiliar people and voices can so easily start that stampede you have been so careful to avoid.

Handling is another activity liable to lead to panic stations, so don't catch and handle the young stock any more than is necessary.

Second-stage rearing

After their first four weeks in the brooder house, table birds will be ready to start the second half of the rearing programme. Again the object will be to provide an environment in which they will prosper with the minimum of checks. At four weeks of age the table duck that is reared indoors will be nearly halfway to its final weight at nine weeks of about 7 lb. The duckling which is reared partly outdoors will take another couple of weeks to reach 7 lb, provided disease, predators and vermin are kept at a distance.

The brooder house can be used for the fattening

stage, but it will help if the birds can be given a change of run. If it is of grass it should be kept short so as not to provide a happy hunting ground for parasites. A straw yard, as discussed in the next chapter, is another alternative.

As during the rearing period, keep management simple and regular. If it is possible to adopt a routine whereby the birds are let out, fed, watered and shut up at the same time each day, they will respond that much better.

In the second stage it is best to keep the feed troughs and drinkers in the pen outside the house, but if this system attracts too many wild birds, place the trough just inside the house and the drinker just outside so the ducklings do not have far to travel to wash their beaks. If they get a feed just after shutting-up time, make it a wetter mix to compensate for the lack of water overnight.

Drinkers are best left outside because ducks are positive geniuses at splashing it all over the place. It does not matter in the open where the water will dry (although it will help to mount the drinker on a wire base over a shallow pit to give drainage), but a drinker in the house will leave a permanently waterlogged area which will quite confound your efforts to provide dry living-quarters.

Rearing to point of lay

The parting of the ways for table ducks and laying ducks comes around nine or ten weeks of age. The table bird goes off to the oven; the layer embarks on the final six or seven weeks of rearing before the first eggs start to appear.

As with table birds, the same quarters can be used throughout the rearing period, but there are obvious advantages if you have a second house and run in addition to the brooder house. Replacement ducklings for the next batch can be brooded without disturbing the adult flock and the brooder house itself is not pressed into continuous service.

The first feathering already encountered with table birds marks the laying ducks' emergence from the duckling stage. It is now, so to speak, an adolescent. Its hard feathering makes it able to survive in long grass and to go swimming. The first eggs will start to appear at about four to $4\frac{1}{2}$ months of age, depending on the date the duck was hatched. During the last two months of the growing period, the adolescent layer can be housed in the type of dry, airy but draught-free building described in Chapter 3.

Like all adolescents, the ducks will have tremendous appetites, which will be growing week by week. Remember this and take care to keep pace by the addition of extra trough space. It is in feeding that the first acknowledgement comes that they are indeed adult ducks.

At about $3\frac{1}{2}$ months old, they can be taken off the growers' meal which has been their staple diet up till now and converted, over a two-week period, to layers' rations. It is a bit like the key of the door—once they are on layers' feed full-time, they are regarded as adult stock.

Life of a layer

Nine out of every ten duck eggs are laid at night or in the early morning, so to avoid finding eggs in all parts of the garden or run, keep the flock locked up until 9.30 am at least, and as soon as the ducks have been let out you can go round the house and collect the day's eggs. Frequent collection will check the danger of salmonella in the eggs caused by soiled eggs.

Meal-times can be arranged round this programme with the morning feed given as soon as they are let out and the evening one about an hour before they are shut up for the night. In the summer, this will mean pushing the last feed on into the evening.

Ducks, unlike chickens, do not suffer a check in egg production when the nights start to get longer. Chickens, left to their own devices, will ease up on egg output during the autumn unless their houses are lit. Ducks, bless their webbed feet, press on regardless, with the only check coming when they moult sometime after mid-August. Even then it is a relatively minor check lasting not more than six weeks. First, production will fall gradually to about 50 per cent, or to an egg every other day. Then it will fall more rapidly and for a week or ten days there will be virtually no eggs at all. Once production returns the flock will be back to 50 per cent production within about ten days. A warning: just because there are no or very few eggs coming during the moult, do not look on this as an excuse to relax on the feeding. Quite the reverse, in fact.

Clipping their wings

To ensure that your flock will never take off and fly out of your life and garden, it will be necessary to pinion young birds or clip the wings of adult stock. Pinioning should be done in the first ten days of life with a sharp knife. Take just one wing, rest it on a piece of wood and remove the last $\frac{1}{4}$ inch (see Fig. 4). This is the section which will take the flight feathers in due course. Done early enough in the bird's life, it is a painless operation and will do no harm. But remember, one wing only. The object is to throw the duck off balance. Pinioning both wings would delay flight, but the bird would adjust in due course and be able to take off eventually.

The secret with adult stock which have not been pinioned is to remove parts of the flight feathers, again in one wing. They are the ten or twelve hard, sharp,

Fig. 4 To prevent birds from flying away, one wing should be clipped. If this is done during the first ten days of life, the end $\frac{1}{4}$ in. of the wing can be cut off along the line shown. Use a sharp knife. It is a painless operation for the bird. With older stock trim $2\frac{1}{2}$–3 in. off the ten or so flight feathers with a sharp pair of scissors.

long feathers at the end of the wing. Using a pair of sharp scissors, trim the top $2\frac{1}{2}$–3 in. of the feathers. On no account touch the inner, secondary feathers on the wing which provide welcome protection in the cold weather. Ensure that there is no blood inside the feathers beforehand. This is done by plucking one out; if there is blood inside, it is still growing, so wait a while.

While on mobility, or rather lack of it, a word on driving your birds. Ducks react most readily to anything that appears over their heads. Instead of moving to one side to change their direction, use your arms in the manner of Frankenstein's monster in those old American films. If you want them to go to the left, raise the right arm to shoulder height and the flock will swing in the desired direction. If a right turn is needed, the left arm.

Geese

Hay-box brooding for goslings

Goslings, like all young stock, need supplementary

heat in the early stages. But they are hardier than most and the heat can be turned off after about ten days, except in really bitter weather when extra warmth would be appreciated. As with ducks and turkeys, the same sources of heat are available, but again the most convenient for the small garden unit is either infrared bulbs or a mother hen.

A 250-watt bulb will take up to fifty birds with a similar arrangement of surrounds, feeders and drinkers as with the ducklings (see page 22).

If you are dealing with a handful of six or twelve goslings, an ordinary 100-watt bulb and a 'hay box' will be adequate. Assuming you have six goslings, take a shallow, 3 ft square box (maybe even the box in which they arrived), fill it lightly with hay and place the goslings on the hay. Within a short time they will have pushed it to the four sides so that you have a surround of hay 3–4 in. thick. Once the floor of the box is cleared, cover it with shavings to a depth of about an inch.

Lower the lamp over the goslings, setting the height at just under 15 in. to give the correct amount of heat. Place feeders and drinkers as far from the source of heat as possible, but not so close to the outer wall or surround that they impede access.

Geese are champion splashers and will get water everywhere. If the lamp is too low in the early stages they will flick water on to it and cause it to explode, which shows how important it is to have a spare lamp handy.

Early care and development

Dehydration is a killer of young stock, so make sure there is plenty of water available all the time. The ratio should never be less than one drinker to every twenty goslings. The area around the drinker will inevitably get splashed, but the worst can be avoided by placing it on a piece of wire netting or welded mesh through which the water will drain. As soon as the birds go outside, put the drinker outdoors as well.

If only a few goslings are involved, egg trays or egg boxes will again make excellent feed trays in the early stages.

If the weather is reasonable, goslings can be given their first taste of the open air at a week old. But keep a sharp eye out for heavy rain, as they will not have enough wet-weather protection until fully feathered. Provide overhead protection against predatory rooks and crows. If the brooder house is raised off the ground, provide a ramp down which they can descend.

The permanent switch from brooder house to range can come when the goslings are about three weeks old, or four weeks old in the winter.

As natural grazers, geese will supplement their diet with grass. But this does not mean they do not need feed troughs out on range or that any old piece of pasture will do. Long, coarse, stringy grass is no good; it should be relatively short.

Geese will make short work of any weeds. The dedicated gardener may find the species frustrating when they remove his prize blooms, but he will approve of their weeding habits, particularly in an orchard or among strawberry plants. One vital point— do not allow them access to any land where insecticides have been used.

It is their grazing habits which lead to the production of watery, messy droppings. These in turn mean that the litter floor must be renewed more often or topped up more often than with other species of poultry. A favourite method of littering the floor is a 3–4 in. base of straw, topped with shavings. The straw can remain, but the shavings should be topped up as and when necessary. Normally, the top remains dry because the moisture drains away through the straw.

Like ducks, geese are waterfowl and if you can provide some water for swimming it will be appreciated but is not essential. If a pond is available, or is provided on the lines described on page 45 for ducks, put it out of bounds when the cold weather comes. Loss of body heat to the icy water burns up food which should be making meat.

Turkeys

Four weeks' heat for turkey poults

Turkeys are not the best starters in life and will really need some tender, loving care to get them safely through the first four weeks of life and 'off-heat'. Even the professionals will grudgingly admit to a mortality rate of about 6 per cent in the brooding stage, and as each poult will be costing in the region of £2, you will not want to lose too many.

Like the other types of stock covered in this book, the turkey's basic needs are simple: well-ventilated, but not draughty, quarters; some form of artificial heating in the first weeks; and a clean, disease-free environment. The ventilation requirement means that air should be coming in at the top of the house and not whistling under the door or through cracks in the woodwork.

The 6 ft × 4 ft brooder shed, described on page 20 for ducklings, will serve equally well for young turkeys. But you will need to be a little more particular about floor coverings. Wood shavings are preferable, but they must not be dusty or mouldy and a likely source of disease. Dusty litter is a common source of eye trouble in poults. Hay should never be used, but the shavings can be topped up with straw after two or three weeks.

Whatever the material, the litter must be kept dry. Wet litter will harbour blackhead, coccidiosis, bacterial infections and other potential killers. All litter should be removed when the flock finishes in the brooder house and renewed for the next batch of poults.

Four weeks is the minimum time for poults to be under the brooder. In the winter months, brooding time can be stretched to five weeks. Poults also need it hotter in the early stages and target temperatures in the first week should be 100–105°F.

Again, the infra-red system (page 22) is the best brooding method and the bulb should be suspended about 15 in. above the litter and then raised to reduce the temperature in the second week. A cardboard surround will prevent the poults wandering too far.

If you are rearing about half-a-dozen poults, the hay-box system described for goslings (page 33) will apply equally well to poults with the added refinement that the box should be taken into the range along with the birds at four weeks of age (five weeks in winter). Leave the brooder with them for a week so so that they can use it when they feel in need of its warmth and familiarity.

Early care

Turkey poults are not good foragers. They have to be virtually led to the trough and the drinker before they begin to feed. Details of how to overcome this problem are given in the chapter on feeding (see page 72).

Simple, shallow improvised feeders will get the poults through the first week or so, but take particular care to see that they are clean. Egg boxes or trays are ideal for they prevent the poults lying in the food, but as a disease precaution use new trays only. It is worth spending a few pence on boxes or trays and throwing them away when the poults move on to troughs.

The first drinkers should also be sufficiently shallow to prevent the poults wetting their bodies. Initially, they will paddle, but they should not be able to immerse themselves any further. If adult drinkers are used for young stock, put in a bed of clean pebbles or marbles to lift the water to the lip.

Again, the feeders and drinkers should be beyond the range of the heat bulb, but this does not mean they should be in dark corners. On the contrary, good lighting is needed around them to ensure they will be seen by all the poults, 24 hours a day. In the first three or four days, make sure the poults have enough eating time.

Adopt the 'getting to know you' routine already described for ducks (page 28) and give the poults warning of your approach. Talk, hum, whistle or sing to them as you carry out the different chores and keep visitors away in the early days.

Do not be too disconcerted if the new arrivals appear to walk around in a daze when they first arrive. They may look as though they are partly blind and go into a crouch at an unfamiliar sound or movement, but these are inborn reactions which stem from their ancestors' days in the jungle. The habit will fade as the poults become familiar with the stockman.

If poults are floor-reared, as opposed to in the hay-box, the area enclosed by the cardboard surround can be extended at five days and removed at ten days, when the light intensity should be reduced.

After the first four days on continuous light, a sixteen-hour day should be adopted and at ten days the light intensity halved—say from 100 watts to 50 watts. Poults will react to too bright a light by pecking each others' feathers or vents and if blood spots appear, the practice will intensify.

Once they get the run of the whole floor it will be necessary to prevent them crowding in the corners and trampling and suffocating each other. A truss of baled straw, or some of the corrugated cardboard surround, placed in the corners will solve this problem.

The sign of a satisfied flock in these early stages is of energetic poults, hurrying all over the place busily seeking food and happy with life. If they are huddling in groups, the temperature is too cold; if they form a ring beyond the range of the brooder, they are too hot and the heat bulb should be raised. Avoid these two extremes and make supplies of feed and water constantly available and all should be well.

Obviously, if you buy-in poults at four to five weeks, much of the foregoing can be ignored. Remember instead that the birds' need for a drink will be greater than that for feed and keep the lights on at night for the first day or two until they are settled in.

3 Shelter Against the Elements

Housing ducks

Water may run off a duck's back, but that doesn't mean that she wants to spend her nights in a place where the roof leaks and the wind whistles round her beak.

Ducks do not seek anything elaborate in the way of sleeping quarters. All they ask for is somewhere that is rain-proof, wind-proof (but not stuffy), and proofed against rats and other predators.

To fulfil the first two requirements, siting is a vital factor. The building should not be in a hollow into which water drains from surrounding higher land. A better position is on slightly sloping ground which will drain freely. If there are any small hollows within the ducks' run they might prove a problem, so dig them out and fill with stones to a depth of 6–8 in. The house also needs to be away from the shade of tall trees and hedges which will keep the sun off and prevent the site from ever drying out completely. Ducks need shelter from incessant sun, but their quarters need the benefit of the sun's drying qualities.

Windows and doors should face away from the prevailing wind and roofs could have gutters and down-pipes to take away rain water.

Inevitably, the area immediately outside the house is going to be well paddled, and a soakaway will help keep the run sweet and the ducks relatively clean. Although ducks like water, they abhor damp; it can bring on rheumatism to which they are prone.

A roof and three walls

Having got the house in the right place, let us look at the type of building suitable for a small flock being kept either for eggs or meat. With the addition of some method of keeping the house warm, it will also serve as a brooder house for young stock.

Obviously, the house needs a good roof to keep out the rain and for this the well-tried combination of tongued-and-grooved, $\frac{3}{4}$ in. boarding and a covering of roofing felt takes some beating.

Corrugated asbestos is an alternative, but avoid see-through plastic roofing. Ducks can be nervous creatures at night and are liable to start quacking at a stray light, a trait which will not be popular with your neighbours if it goes unchecked.

Three-quarter inch boarding will also serve for the walls. The boarding can be nailed or screwed to a 2 in. × 2 in. frame and need cover just three of the four sides. The door and air inlets will take care of the fourth walls.

There are cheaper ways of building the walls and among the alternatives are straw bales supported by wire netting or two walls of wire-netting with bracken or straw stuffed into the cavity. Hessian sacks, three or four layers thick, can also be used. But there is a considerable degree of make do and mend about these compromise methods and they will need patching up and renewing from time to time. Better, if possible, to invest in something more solid if you intend keeping stock for some time.

A firm base

Two points to consider on the choice of floor—rats and bugs. Rats are a constant menace, particularly to young stock, and a house with an earth floor will present them with no problem at all. They will simply dig under the walls, unless precautions are taken.

A concrete floor will have the added advantage of being easy to clean and disinfect. A substantial wooden floor has the same advantages and will be warmer. Concrete will require an extra few inches of litter to make it comfortable for the ducks. If neither concrete nor wood are practical, rats can be kept at a distance by 12–15 in. wide strips of wire-netting buried in the earth around the house, rather on the lines of the submarine nets which protect harbours from attack in time of war.

Another method of keeping rats at bay is to raise the house 12–18 in. on brick or concrete piers, with anti-vermin collars fitted halfway up. In addition to making life harder for the rat by cutting out under-floor hiding places, a raised house will also be a drier house—the air circulating under the floor helps dry any damp patches. A wooden walkway will be needed for the ducks to descend when they emerge in the morning—they will not relish starting the day with an 18 in. drop!

Floor coverings

There is a wide choice of litter materials: peat moss,

wood shavings, chopped straw, sawdust and bracken can all be used to a depth of 3–4 in. Whatever material you choose, make sure it is clean, sweet and free from mould.

There are two schools of thought on litter management. One says the litter should be changed regularly, particularly where the birds sleep and when there are signs of dampness. Moulds flourish in damp litter and can affect the quality of the eggs. The second view is that litter should be left and damp patches sprinkled with fresh material which will soak up any excess moisture. Over a year or so, the depth will build up to 12 in. or more, so providing the ducks with a warm, comfortable bed as the material gently ferments. When the house is finally cleared, the litter, with its year's supply of droppings, will prove a useful garden fertilizer.

A final point about building up the litter: the door will need a threshold to hold back the increasing depth. Without it, the peat, shavings or sawdust will spill out into the run.

Room to move

In the house, each adult member of the flock needs around 3–4 sq. ft of floor space. A house measuring 8 ft × 6 ft, as shown in Fig. 5, will take between twelve and fifteen laying ducks. Ducklings being reared for their meat will need less room in the first four weeks, but they grow rapidly and their space requirements expand at a similar rate in the last four weeks of life.

Err on the generous side for adult stock. Ducks are temperamental creatures and subject to sudden moments of panic. If a car backfires or a door opens without warning, one or two members of the flock will start a quacking which will quickly spread to the others. In their panic, they will bunch into a group

Fig. 5 This 8 ft × 6 ft house will take up to fifteen adult ducks or eight to ten geese. The door, at about 2 ft 6 in., is wide enough to allow ducks to emerge in the morning without falling over each other. The wire netting ventilators under the eaves will keep the interior airy but not draughty. At night they can be covered with shutters, which turn down during the day.

and if the house is too tightly stocked some will get damaged in the crush. Remember this tendency when planning a door for the house. A pop-hole which allows one bird through at a time will not be sufficient if it is the only method of getting in or out of the house.

Ducks don't have the patience of chickens and when the house is opened in the morning, they will be jostling to be first out into the run. If the door is too narrow there will be a traffic jam; the ones at the back will be climbing over those in front and some will inevitably get hurt, so give the birds plenty of wing room on the way out. Four feet would not be too wide for a flock of thirty or more. If a door of that width is cumbersome to open, it can be divided in two and double doors installed.

The doors should open outwards so they can be used to control the movement of ducks in and out of the house. Assuming a 2 ft wide door, the first post of the netting for the run can be placed 2 ft away. When the door is opened at right angles to the house, it latches to the post so that when the ducks are driven back at night, they come against a wooden wall. They have no choice but to turn in the opposite direction away from the wire of the pen into the house. This system completes the run and avoids the need for a second door in the wire.

Let there be light – but not too much

Remembering again the ducks' propensity to panic, avoid windows as these will let in stray shafts of light. Go instead for wire-covered air inlets fitted just below eaves height at the front with an additional one at the side of the house. The wire will keep out stray birds and foxes.

The great outdoors

The ducks will spend only the hours of darkness in the house. The rest of the time they will be out on the range or, as is more likely with a garden, in a wired-off pen.

The commercial industry talks in terms of 100 ducks to the acre, but a more realistic figure for domestic use is about 100 sq. ft for six ducks or 150 sq. ft for twelve ducks. Obviously, the pen is going to come in for very heavy wear during the season so, to avoid it becoming a mud patch, two pens should be made available if this is possible. A compromise is a smaller concreted area for use in wet weather, leaving the grass run for the dry days (see Fig. 6).

The netting round the pen should be of 2 in. mesh, staked every 10 ft or so. Don't be tempted to go for a cheaper, wider mesh. Apart from letting ducklings through, anything wider than 2 in. will be difficult to handle and will require more staking.

If the birds' wings are clipped (see page 31), netting need only be 18 in. high. For the lighter breeds,

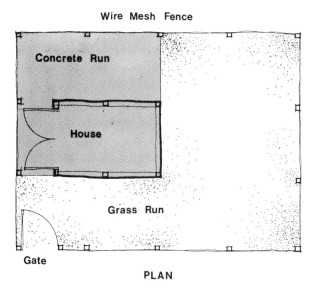

Wire Mesh Fence

Concrete Run

House

Grass Run

Gate

PLAN

Fig. 6 A suggested layout for a duck house with a choice of run. The concrete paving can be used for wet weather and when the grass is being rested. Double doors across one end of the house can be opened accordingly to which run is being used.

whose wings should also be clipped, the height of the wire should be increased by about 1 ft. If possible, arrange the pen so that one side of the house forms one of the sides of the pen.

Straw yards, widely used by the commercial industry, can be an alternative system for the garden (see Fig. 7). Table birds will thrive in a low house built on these lines. It opens on to a wired pen covered with a 6 in. layer of straw, which can be topped up as it gets soiled. A straw yard can be stocked more tightly and one of 100 sq. ft could take up to twelve ducks. Floor space inside the house can come down to 3 sq. ft a bird.

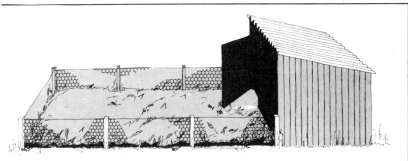

Fig. 7 This type of duck building opens out on to a straw yard bordered by a wire mesh fence. A 100 sq. ft straw yard will take up to twelve ducks with an allowance of 3 sq. ft a bird inside the house.

Water for swimming

Ducks are water birds, of course, but they can live without water for swimming. Although it is a splendid sight to see them swimming and splashing in what seems a natural environment, a pond is not a vital component of a garden duck unit unless you intend to breed from a heavy strain, such as the Aylesbury. When ducks get beyond average size it is difficult for male and female to get together anywhere but on water.

But don't be put off if your birds are average size and you would still like a pond. Observe a few elementary rules about keeping it clean, be prepared for a few extra chores and a pond will add character.

If the garden has a stream running through it, there is no problem, but if, as is more likely, there is no natural waterway you will have to think in terms of manufacturing a pond. Those ready-made pre-cast ponds will be quite adequate, but an old galvanised cistern, sunk into the ground, will be cheaper, as will a sheet of thick polythene placed in a hollow.

The water will have to be drained and cleaned

periodically and a drain hole, suitably plugged, will have to be included at the base. The best scheme is to run a pipe from the hole to carry the water to a soak-away. The pond can be filled by using a garden hose or permanently laid pipes. If you haven't either, the only alternative is a chain of willing volunteers and a good collection of buckets.

Housing geese

Geese, like ducks, are prone to cramp and rheuma-tism and their accommodation requirements are similar. All they look for is a comfortable bed, a dry floor and a roof which will keep out rain and protect them from draughts.

You will soon know if the house is not to their liking for they will not enter when driven in at night; one of the first signs of damp litter or a draughty floor is that they behave like naughty children when it's time to go to bed. If this happens, check the condition of their sleeping quarters.

The duck house shown in Fig. 5 will do equally well for geese except that the number of birds will have to be reduced. Allow $4\frac{1}{2}$–5 ft per bird, so the 6 ft × 8 ft house will hold eight to ten geese. Again, the floor can be concrete or hard-packed soil with the litter, i.e. shavings, peat moss, etc., to a depth of 3–4 in.

Geese, as we have seen, are grazing birds and they will continue to graze after dark. It is not unknown for them to be left out at night when required to put on as much weight in as short a time as possible. If they are left out, pay particular attention to the defences against foxes, an ancient enemy of geese.

Protection will be needed wherever foxes are particularly active, and these days this can often mean built-up urban areas. It should take the form of a solid house and a 7 ft high, wire fence that extends 1 ft underground to deter would-be tunnellers.

A portable, apex type of building makes a useful alternative to fixed accommodation, particularly if you have a large garden (at least $\frac{1}{4}$ acre) over which the geese can range. Equipped with carrying handles at each end, the apex house can be moved to a fresh spot every couple of weeks. Suggested measurements for the one shown in Fig. 8 are 6 ft long × $4\frac{1}{2}$ ft wide, with a ridge height of 4–5 ft, giving a capacity for five birds.

The door at the front end will give you a way in and a $1\frac{1}{2}$ ft deep, hinged flap across the back will provide an entrance for the geese. But both you and the birds will have to lower your heads as you enter.

Materials should consist of boarding for the ends, door and flap, and roofing felt and battens, or better still, tongue-and-groove boarding, for the sloping sides. Wire netting will fill the triangle just below the ridge which can be of wood or galvanised iron.

If you have a spare acre of good quality grass, up to 100 geese can be reared, but this will take a high degree of skilled management. In the event, it is likely that

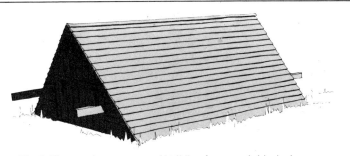

Fig. 8 The popular apex type of building for geese is ideal where birds can be ranged over $\frac{1}{4}$ acre. It can be lifted by the carrying handles and moved periodically to a fresh patch of grass. The geese will go in and out under the 18-in. flap at this end. There is a door for the stockman at the far end. At 6 ft × $4\frac{1}{2}$ ft, the house will take five birds. Ridge height is 4 ft.

you will have less grass available—$\frac{1}{4}$ acre at the most. In this case, the flock should be reduced to about twelve birds and there will be enough grass available to keep them fed during the growing season (see page 69).

Housing turkeys

The most popular form of housing for the turkey in the commercial industry is the pole barn—a simple construction enclosed on three sides with wire netting forming the fourth. There is no reason why this design should not be adapted for a modest garden unit.

The fourth side can open out on to a pen or be wired in with a door opening on to a pen. The house should be sited away from prevailing winds and on the driest, best draining land available. An overhanging roof on the open side would help to keep out driving rain or snow.

The 8 ft × 4 ft house shown in Fig. 9 will hold six birds of seventeen weeks and above, when their space requirements are about 5 sq. ft each. Leading up to

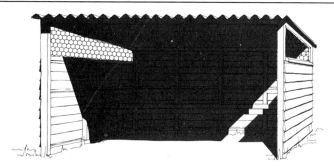

Fig. 9 The garden version of the pole barn which has served the commercial turkey industry so well. This 8 ft × 4 ft house will hold six birds of seventeen weeks and over. The front can be either left open leading on to a run or wired in with an access door. Perches will relieve pressure on floor space.

that stage, normal space allowances are: 2 sq. ft from eight to twelve weeks, $3\frac{1}{2}$ sq. ft at thirteen to sixteen weeks, and $4\frac{1}{2}$–5 sq. ft at seventeen weeks and beyond. Avoid overcrowding, which will lead to cannibalism and wet litter.

A free-draining rubble or clinker base will help keep the litter dry. A concrete or wooden floor will be the easiest to clean between crops and it will deter burrowing predators.

Perches will relieve the pressure on floor space and provide somewhere for the turkeys to roost. Make them of 3 in. poles to support the larger birds, allowing about 2 ft perching space for each bird. The perches should be smooth to counteract breast blisters, but avoid completely round poles as the birds will then have difficulty maintaining a grip; it's better to have square lengths of wood with the corners rounded off. Keep the perches at one level or the entire flock will only use the top one. Birds hate to feel that another member of the flock might be one up on them.

Once they are off heat, turkeys will stand up to most weather conditions except extreme heat, which will dry them out rapidly. Their shelter should not become an oven in hot weather, so make provision for ventilation by leaving a 6–9 in. gap along the top of the walls, just below eaves height. This will keep air flowing, even on the hottest days.

Best Birds for the Job

Duck strains

Unlike the chicken industry, duck people have not

become enmeshed in a mass of hybrid birds that have been developed out of the pure breeds. Perhaps it is one of the reasons why, as a nation, we eat over 300 million chickens a year but only about one million ducks. With two or three exceptions, the most prominent and popular birds are the pure breeds.

Obviously, the first choice to make is: table or eggs? Are you interested in the meat your duck puts on or in the eggs it lays? The answer will govern your choice of bird as the best breeds for putting on weight are not the most prolific layers.

Meat-making ducks

Aylesbury Fig. 10. As the name implies, it originated from the county town of Buckinghamshire, though no one seems to know quite when or how. It is the best known of all breeds and it is invariably a rubber or plastic Aylesbury which is such a popular toy at bath time.

Adult Aylesbury drakes weigh 10 lb and the ducks 9 lb. As a layer it is not in the race (about 100 eggs in a laying year), but as a meat producer it is in the super-duck class with a good proportion of flesh to bone. If you want to breed from Aylesburys, the mating ratio is one drake to four, and never more than five, ducks. Heavy drakes are liable to rupture when mating, so choose the lighter ones.

Most of today's Aylesburys contain some Pekin blood (see below) to improve egg production.

Plumage : white. *Legs :* short, sturdy and orange. *Bill :* flesh-pink colour.

Pekin Fig. 11. Again the name gives its origin away, although it has been bred most successfully and prolifically in the USA since the first importation from China in 1873. In the States it is still the basic table duckling. It is smaller than the Aylesbury, 9 lb for drakes and 8 lb for ducks, but egg output is better at

Fig. 10 Aylesbury. The champion meat-maker, with drakes reaching 10 lb and ducks 9 lb.

Fig. 11 Pekin. Smaller than the Aylesbury, 9 lb and 8 lb, but egg output is better.

Fig. 12 Rouen. Takes time to reach 10 lb and 9 lb, but the meat has a flavour all its own.

Fig. 13 Khaki Campbell. Champion layer, regularly producing 300 eggs in a year.

110–130 a year. Fertility is higher than the Aylesbury and one drake can cope with six ducks.

Plumage : creamy white with yellow flesh. *Bill and legs :* orange.

Rouen Fig. 12. If you are seeking a reputation for the flavour of your ducks, the Rouen is your bird. As you might gather, the breed is of French origin and apart from flavour is noted for its good looks. Both sexes resemble the Mallard, and drakes have the same habit of moulting in October and taking on a distinctive feathering.

A major snag with Rouens is a relative reluctance to mature and the breed takes about twenty weeks before it is ready for eating; the drake reaches 10 lb and the duck 9 lb. It lays less than 100 eggs a year and fertility is such that one drake can manage three ducks only. Like the Aylesbury, heavyweight Rouens are liable to rupture when mating so pick out the lighter weights for this job.

Plumage (drakes) : white and black. *Bill :* green-yellow. *Legs :* terra-cotta. *Plumage (ducks) :* like female Mallards that have had a respray—the brown is deeper and the outline of the feathers more distinctive. Wing bars are bright blue with a strip of white on either side. *Legs :* orange-brown. *Bill :* bright yellow with a black saddle at the base.

Welsh Harlequin This is another breed which takes its time to reach table weight—about sixteen weeks for the drakes to get to 6 lb and the ducks to 5 lb. It has the double advantage that it will achieve its table weight on cheaper feed and kitchen scraps and that the duck will lay up to 300 eggs a year. A placid docile bird, it is less likely to be put off lay by shocks or surprises.

This egg production gives a clue as to the origin of the Welsh Harlequin—out of two sorts of Khaki Campbell stock.

Plumage (drakes): similar to the Mallard. *Legs:* bright

orange. *Bill:* green-khaki. *Plumage (ducks):* mainly cream with a fawn head and neck. *Legs:* dark brown. *Bill:* gunmetal.

The Welsh Harlequin drake has been crossed with an Aylesbury duck to produce one of the duck industry's major hybrids—the Whalesbury. Another dual-purpose strain, the drakes weigh 7 lb at sixteen weeks, the ducks $5\frac{1}{2}$–6 lb. Egg output can be as high as 290 a year.

Egg-laying ducks

Campbell The title of champion layer among layer ducks must go to the Campbell clan, bred out of Fawn and White Runner and Mallard, plus some Rouen blood later. The Campbell frequently comes close to laying an egg a day—300 eggs a year from each bird is the par for the course.

The most popular member of the clan is the Khaki Campbell, Fig. 13. *Plumage (drakes):* khaki with lighter shade on the underparts. Head, neck, stern and wing bars are bronze with a greenish sheen. *Bill:* green. *Legs and feet:* dark orange. *Plumage (ducks):* khaki with a darkening on the wings and neck. *Bill:* greenish-blue. *Legs:* same colour as the body. Adult drakes weigh 5–$5\frac{1}{2}$ lb, ducks $4\frac{1}{2}$–5 lb.

Two other Campbell strains are the White and Dark. In the White, feathering of both ducks and drakes is white throughout, grey-blue eyes, orange bill, legs and feet.

When the Dark Campbell is crossed with Khaki females, the sexes of the offspring can be distinguished by their colouring.

Plumage (ducks): dark brown head and neck, light brown shoulders, breast and flanks, each feather outlined in dark brown. Purple-green wing bar. *Bill:* slate-brown. *Legs and feet:* dark brown. *Eyes:* brown. *Plumage (drakes):* as duck, but with green head and

Fig. 14 Indian Runner. A great survivor with an egg output of 140–180. It has no shoulders.

Fig. 15 Embden. Heavyweight goose champion with top weights of 34 lb and 22 lb.

Fig. 16 Toulouse. Second in the weight league and frequently crossed with the Embden.

Fig. 17 Chinese. Early maturing— can be killed at eight weeks. Top weight 10–12 lb without much fat.

neck. Wing bar is more vivid than on the female.

The White Campbell is as prolific a layer as the Khaki and a popular cross in the production of dual-purpose strains. All colours of Campbell lay white eggs.

Indian Runner Fig. 14. This was the top egg layer until the Campbell appeared. Even though an output of 140 to 180 eggs a year cannot compete, the breed is popular for its ability to survive. They are hardy in the extreme and will thrive almost anywhere. Great foragers, they are ideal for an orchard where they will search out the slugs and insects. Not the most athletic of birds (they will not need a high fence to keep them in), but they do appreciate somewhere in which to swim and splash.

Indian Runners can be distinguished by a complete absence of shoulders and by their upright stance. From a distance they look like bottles of hock on legs.

There are a number of strains in the breed including White, Fawn, and Fawn-and-White. Colouring of the Fawn is: *Plumage (ducks)* : head, neck and keel a light shade of fawn. *Bill, legs and feet*: very dark brown which appears black. *Plumage (drakes)* : head and neck bronze and green; breast and neck base reddish-brown; back dark brown and keel a fawn-green; rump and tail dark brown. *Bill*: black or dark green. *Legs and feet*: black-brown.

All types of Indian Runner ducks weigh 3–4½ lb, drakes 3½–5 lb.

Geese breeds

As it is with ducks, so it is with geese. The hybridisation of the twentieth century has passed the species by. With one or two exceptions, geese are still putting on weight and laying eggs at much the same rate as they were nearly a century ago. One of the few significant breeding advances in recent years has been to produce a bird that exudes less fat when it reaches the oven.

There has been a high level of cross-breeding over the years and many of the geese that waddle across the farmyard are not pure breeds but first crosses, with the Embden-Toulouse combination as the most popular. Both these breeds come top of any league table based on weight: over nine months they will reach 20 lb. Further down the weight table is the Brecon Buff (16–19 lb), the White Chinese (10–12 lb) and the Roman (8–14 lb). The smaller birds also make the best layers and the lightweight Roman will produce 40–45 eggs in a season, the Chinese 40–60, the Brecon Buff 30–40, the Toulouse 30–38 and the Embden 30–38.

Size also has a bearing on mating ratios. In the heavyweight Embden and Toulouse, it is one gander to three geese, but it can increase to one to four or five for the Chinese, Brecon and Roman. If you are producing goose eggs for hatching it is obviously economical to have as few ganders as possible.

Embden Fig. 15. This is the largest of the pure breeds, with ganders frequently weighing up to 30–34 lb and the geese peaking at 20–22 lb. To reach these weights, however, they should not be mated until they are at least two years old, with three females to each male. Despite the amount of fat it produces, the Embden gets high marks as a table bird with well-flavoured, top-quality flesh.

Plumage : spotless white with close-fitting feathers.
Bill, legs and feet : bright orange.

Toulouse Fig. 16. Not quite so large as the Embden, the ganders reach 28–30 lb and the geese 20–22 lb. The females do not have a highly developed maternal instinct and are inclined to leave their offspring to fend for themselves.

Plumage : dark grey head and upper parts, grey breast

with a lighter shade on the thighs; wing, back and thigh feathers edged with white; pure white under the stern and tail, apart from a broad band at the centre of tail. *Bill, legs and feet :* orange.

Brecon Buff Medium weight range with ganders at 19 lb and geese at 16 lb. Better layers than the heavier birds and one of the quietest of breeds; this can be an advantage if you have neighbours who might object to cackling.

Plumage : similar to the Toulouse except that the grey parts of the larger bird are buff in the Brecon. *Bill, legs and feet :* pink.

Chinese Fig. 17. The earliest of breeds to mature, the Chinese can be killed for the table when they are eight weeks of age. At top weight they are little heavier than a large duck—12 lb for the male, 10 lb for the female.

Probably because of its early maturity, the Chinese does not build up so much fat as the larger breeds. Its dark flesh has a similar flavour to the wild or Canadian goose, but its dense covering of down can make plucking difficult. This apart, the breed is ideal for the back garden as it will graze the grass, provide up to sixty eggs as well as meat, and make a first-class watch dog.

The Chinese comes in two colours—White and Brown. The White strain, with its long neck and orange bill, feet and legs, will be mistaken more readily for a swan than any other goose.

Plumage (Brown) : varying shades of brown starting with a dark stripe which runs down the back of the neck from the head to almost the tail; cheeks, throat, front of neck and breast are fawn coloured; brown feathering on the wings; thighs edged white. *Bill, legs and feet :* mainly orange but can vary.

Roman Another small, compact strain with 15 lb ganders and 13 lb females. Egg output is less than the Chinese, at about 45 a season.

Table 1 Weight and food conversion ratios for leading turkey strains

Breeding Company	Type		Live-weight (kg) at (weeks)				Food Conversion Ratio at (weeks)			
			12	16	18	20	12	16	18	20
Attleborough Poultry Farms	Small White	Male		6.12	6.80	7.93				
		Female		4.08	4.76	5.21				
	Wold Gold (small)	Male		6.35	7.48	8.39				
		Female		4.30	4.78	5.44				
	Wold (medium)	Male		6.50	7.92	8.80				
		Female		4.63	5.21	5.89				
	Super medium	Male		7.48	8.04	10.58				
		Female		5.66	6.35	7.25				
British United Turkeys	BUT 8 Heavy Med	Male	7.1	10.6	12.3	14.0	2.13	2.49	2.69	2.91
		Female	5.4	7.5	8.2	8.8	2.36	2.87	3.17	3.51
	BUT Big 6 Heavy	Male	7.94	11.99	13.9	15.9	2.11	2.44	2.64	2.85
		Female	6.11	8.76	9.88	10.8	2.29	2.75	3.00	3.28
Kelly Turkeys	Super Mini	Male		6.12	7.48	7.93		2.7	3.0	3.2
		Female		4.32	4.95	5.45		3.1	3.5	3.8
	Wrolstad	Male		7.12	8.25	9.21		2.7	2.9	3.1
		Female		4.99	5.62	6.21		3.0	3.4	3.6

Breeding Company	Type		Live-weight (kg) at (weeks)				Food Conversion Ratio at (weeks)			
			12	16	18	20	12	16	18	20
Leacroft Turkeys	Leacrofter	Male		6.6	7.6			2.68	2.9	
		Female		4.68	5.3			2.93	3.17	
Nicholas Europa Ltd	White commercial	Male	7.33	10.88	12.82	14.79	1.92	2.36	2.57	2.75
		Female	5.62	7.8	9.00	10.07	1.97	2.47	2.70	2.97
Ross Poultry (Great Britain) Ltd	Ross Traditional	Male	4.70	7.13	8.20	9.26	2.26	2.66	2.88	3.13
		Female	3.46	5.00	5.60	6.13	2.48	2.97	3.27	3.60
SCF (Turkeys) Ltd	Wirral White	Male	4.75	7.1	8.25	9.37	2.2	2.6	2.8	3.1
		Female	3.64	5.05	5.59	5.97	2.5	3.0	3.3	3.6
Edward Webster Ltd	Bronze 1	Male	5.1	8.0	9.2	10.5	2.28	2.66	2.87	3.12
		Female	3.9	5.7	6.3	6.7	2.45	3.00	3.29	3.69
	Plumpie	Male	5.7	8.7	10.1	11.4	2.13	2.55	2.76	3.03
		Female	4.4	6.3	7.1	7.8	2.45	2.98	3.30	3.70
	Roly Poly	Male	4.7	7.0	8.2	9.3	2.35	2.74	2.96	3.21
		Female	3.5	5.2	5.8	6.2	2.61	3.15	3.46	3.88

Again, a large build-up of fat is avoided and in addition to providing a tasty carcase the Roman is a hardy, active and fertile breed. One theory is that the Romans can be killed when they are in first full feather at about eight weeks. Although little more than goslings, the birds are said to be good to eat and economic to produce at this tender age.

Plumage : white. *Bill, feet and legs :* orange-pink.

These smaller breeds are being increasingly used in cross-breeding. The Roman female is often crossed with the Embden to give better egg production and to increase growth rates. White Chinese are being used in much the same way and research work has indicated the value of the three-way cross with, for example, an Embden gander being put to an Embden × White Chinese or to a Roman × White Chinese goose.

Turkey strains

There was a time when basic choice of stock in the turkey industry rested between the Broad Breasted White, or Broad Breasted Bronze. As the industry developed, however, the old strains gave way to commercial hybrids all designed to produce the right shape of bird in the shortest space of time and on the minimum amount of feed. The modern turkey is a major masterpiece of genetic engineering and each poult which comes out of an incubator represents years of research and development.

Each strain is designed for a particular job. There are heavyweight strains where the stags reach about 30 lb at twenty-four weeks; the mediums are killed earlier at sixteen weeks, weighing around 16 lb for the stags and 11 lb for the hens; and a number of mini-strains are available which, in about twelve weeks, reach about 9 lb for the stags and $7\frac{1}{2}$ lb for the hens.

Feed conversions for all the three sizes is in the

region of 2·5 to 3·0. This means that a hen with a conversion ratio of 2·5 : 1 will turn $2\frac{1}{2}$ lb of feed into 1 lb of turkey. Details of weights and feed conversions of the leading strains of turkey available in this country are given in Table 1. These figures, provided by the breeding companies, are indications of what can be achieved on commercial units. There is no reason why they cannot be achieved in a garden unit. Full addresses of the breeding companies are given in Appendix A, page 108.

Feed From the Bag and the Kitchen

With all livestock—and ducks, geese and turkeys are no exception—the better the quality of feed you give them, the better the product they give you. But whereas geese, as will be explained later, will grow well on plentiful supplies of grass, the other two require large amounts of ready-made feed if the venture is to be worthwhile in economic terms.

Laying ducks, for example, could probably survive quite well on household waste, a bit of cereal and pickings from the garden, but they would not lay many eggs. Neither would there be much of a meal on an eight-week-old duck or even an eighteen-week-old turkey kept on such a diet. Certainly there would be some meat, but the chances are that you would be better off buying a chicken from the butcher for all the money you would save. The problem is where to draw the line.

There are, of course, plenty of ready-mixed feeds

available of the type and quality used by the large commercial companies. But even they have to pay high prices for the right food—anything from £160 to £180 a ton, and this is with substantial discounts for bulk buying. If you were buying feed in the sort of quantities needed for six or a dozen birds, it could cost you double this amount.

To put this in perspective, take the case of a fattened duck. The commercial producer would reckon to produce an 8 lb bird in about eight weeks, using 32 lb of feed in the process. At 7.5p per lb it will have cost him in the region of £2.40 for feed alone, not counting any equipment or the cost of the duckling. So it can be seen that if you go in for the proper feed, at possibly double the price, your 8 lb duckling will cost you something like £4.80 just for feed.

Since the chances are that you are looking more for pleasure of home production rather than for a profit or a particularly cheap meal, this costly approach may not worry you. Certainly there is a lot to be said for it in that it simplifies the job.

There is, however, a compromise. Much of the sting can be taken out of the high cost of feed by making it go further with cheaper ingredients such as kitchen left-overs, bakery waste and the like. The penalty will be less growth or fewer eggs. But then maximum production does not have to be the name of the game; a plentiful supply of home produced food is a fair reward.

In this chapter, both approaches—the high cost and the cheaper method—will be discussed. But it must be made clear that where we are rearing very young stock there is only one approach—second best will not do. The whole success of the job relies on getting the birds off to a good start in life; weaklings only become a burden, if they survive, and are susceptible to diseases and the elements. So for the first few weeks, in the case of fattening ducks, turkeys and geese, or the first $4\frac{1}{2}$ months in the case of laying ducks, we must give ready-made high-quality starter feeds.

General hints

Before we get into the more detailed aspects of feeding, it is worth listing the sort of kitchen wastes suitable. Such items as stale or fresh bread, potatoes, potato peelings (cooked), buns, pies, cakes and pastry, burnt toast, apple peelings and cheese rind are all acceptable, boiled up to make a mushy 'soup' if you wish.

Neighbours, local schools and hospitals are all good sources of waste scraps if you haven't enough from your own kitchen, but you are obliged to get a licence from the local authority before using waste from their property.

There are some items, however, which should never reach the food bucket. Do not give the birds food that is unfit for human consumption. Stale bread, for instance, may not be particularly palatable to us but it is edible. Rotting meat, though, is a different matter. Use your judgement. None of the birds with which we are concerned can tolerate salty food, so plate scrapings contaminated with a lot of salt should therefore be avoided. Similarly, uncooked potato peelings are of little value, as are orange, lemon and grapefruit skins, coffee grounds, tea leaves, banana skins, decomposed meat and fish.

Bones and small quantities of scraps can be boiled up with swedes, carrots and the like to produce a rich nutritious liquor for adding to bran and cereals, a sure way of cheapening the feed (more about this later).

The reason we cannot just feed birds on 'anything' is that over the years they have been bred for a fast growth rate or a high egg production with a low appetite. Household scraps contain a lot of water and therefore fill the birds up without providing the

63

necessary goodness. Hence the need for the concentrated, ready-made foods.

Corn merchants are the best suppliers of feed as they
can furnish you with $\frac{1}{2}$ cwt bags of either ready-mixed
feed or ground cereals or grain. Pet-shops, unless you
can arrange some substantial quantity discounts, are
expensive sources of feed. If you can buy in $\frac{1}{2}$ cwt
bags, store them in a dry place. Plastic dustbins are
suitable for storing unopened bags and a few of these
around the place will help considerably, as will a decent
light-weight shovel and something in which to mix the
feed.

Feeding laying ducks

Rearing

The aim with rearing prospective laying ducks is to
build up their body reserves so that they come into
lay at four to four-and-a-half months, well prepared
for a long, productive life. As already mentioned, a
good quality compounded ration is advisable for the
first three to four weeks. There will always be someone who gets away with feeding hard-boiled eggs,
baked bread, milk and biscuit meal, but this is not for
the inexperienced.

First five weeks

In most instances, specially prepared duck foods will
be difficult to obtain, but chicken feed is just as good.
What you need is a chick starter ration, available either
as meal (mash), crumbs or pellets. There is no real
difference in their feed value, but pellets are the most
convenient, followed by crumbs. If meal is fed, it
should be dampened to a crumbly, moist consistency.

The pellets or crumbs should be made freely
available at all times, allowing the youngsters to help
themselves from waste-proof troughs. Some recom-

mend starting ducklings on bread and milk, which is worth trying, but pellets should still be available.

If dampened mash is being fed, it should be offered about four times a day in quantities that are eaten up in about an hour. The first meal should be given as early as possible in the morning and the last one just before locking up.

During this initial growing stage, some chopped green food will be much appreciated. Even more vital is the presence of drinking water—day in, day out; on no account should the birds be without fresh water.

Grit is an essential part of the diet in that it lodges in the gizzard and helps digestion by grinding the food and it should be fed after the first week. Oyster shell, for bone formation and later egg shells, should be provided after two weeks. Again, this can be obtained from merchants or pet shops; a handful should be sufficient for a dozen birds. It is vital to ensure that there is always some of both types available. Initially only fine grit should be used, which can be increased in size as the birds grow.

Five to twelve weeks

Again the most simple way of feeding growing ducks is to provide chick grower rations either as mash or pellets if available. If mash, it can again be dampened and fed three to four times a day, the amount being as much as the birds can consume in an hour, or it may be fed dry. At this stage you can substitute mash made from a mixture of barley meal, bran, ground oats, cooked maize and a little fish meal and bone meal. This can be dampened with hot water or hot 'soup' prepared from a bone and scrap mixture as described on page 63. It should be made crumbly and moist and given as about four meals a day, again the amount being as much as the birds can consume in about an hour.

By about eight weeks of age the feeds can be cut

down to three times a day, then twice a day by ten weeks, with perhaps a light meal at mid-day. As an alternative, one of the meals can be replaced with un-crushed wheat, barley or oats or a mixture of all three, from eight weeks of age.

Into lay
Depending of the hatching date, birds will come into lay at about sixteen to twenty weeks of age, so we should be building towards this event from about twelve weeks onwards. Again, if duck feed is not available, layers' mash is suitable. It is best, at least until the birds are well established in lay, to feed this undiluted as a dry mash offered fresh twice a day. Around point-of-lay, ducks will be consuming 6–7 oz of feed a day each, so this can be used as a guide as to how much feed to offer at each meal. And don't forget the oyster shell and flint grits.

Later on in lay we can think in terms of cheapening the ration again, but avoid a sudden change in the feed. Introduce new feeds gradually over a period of a few weeks.

The home egg producer can make use of baked, pounded bread scraps, minced cooked meat and fish scraps, mashed potatoes and peelings, grated raw carrots, onion and onion tops, carrot tops, beetroot, cooked parsnips, mangolds, scrapings from cooking pots, cheese and non-salty bacon rinds. You can also include green foods like dandelion, lettuce, kale, cab-bage and other allotment surplus and thinnings. If bread is available in large quantities it can be substi-tuted for the mash virtually pound for pound.

Cook any or all these ingredients with the minimum amount of water and mix it with a special layers' mash at the rate of 1 lb of mash to 3 lb of cooked scraps and greens. If there is not much meat or fish in the scraps, add a little fish meal or meat and bone meal, and a teaspoon of cod-liver oil. Again, it should be as a crumbly mash and the daily allowance should be about 10–12 oz per duck, given as two meals. The mixture can be boosted with small amounts of dried yeast, grass meal and dried kitchen waste.

If ducks have a decent run, the essential animal-protein part of their ration will be augmented by natural foods, such as grubs and flies. Unfortunately, ducks tend to moult in June to July when the ground is dry and grubs are in short supply, so it is wise to boost the protein food in the mashes with fish meal and meat and bone meal either bought-in or prepared at home from scraps. They can all be pressure cooked and added to the mash. Prolonged frosts will destroy grub life and supplies will be limited anyway if the ducks have only a small grass run.

Generally speaking, the food trough should be as near the water trough as possible since ducks like to wash their bills while feeding. Feed twice a day, although at certain times of the year when natural food is plentiful, you might find that the morning feed is not much appreciated. If this is the case, instead of leaving food around to go stale, feed less in the mornings and give as much as they want in the evenings.

The general theory in feeding is to get a balance between the energy, protein, fibre and vitamins in the diet, which is why we feed grain (or bread) and fish meal (or fish scraps). Ideally, the ration should be balanced every day although this is not essential providing, over a period, reasonable amounts of protein and vitamins are being fed. To a large extent, ducks will look after their own needs when they can.

Feeding fattening ducks

As with layers, a well-grown meat bird needs a good start in life and this means buying-in proprietary compounded feed. Duck starter crumbs or pellets (containing 10 per cent protein) are the best early feeds, but can safely be replaced by broiler starter pellets or crumbs. When you are buying, insist that the rations do not contain any additives of the type normally included in broiler feeds.

The pellets or crumbs should be freely available, twenty-four hours a day for the first four weeks of life. It would be feasible to feed this way right through to killing, but it is extremely expensive and not really justified. So after four weeks, make a change to a lower protein (therefore cheaper) growers' mash.

To cheapen the feeding even further you can dilute the growers' mash with a combination of the boiled scraps and vegetables mentioned earlier. A mixture of 'kitchen-scrap soup' with bran, ground oats, barley meal and perhaps a little maize meal will provide a reasonable fattening diet. It can be taken even further, particularly if the ducks have a good-sized run with plenty of natural food.

A mixture of soaked stale bread or confectionary waste with some boiled fish scraps will make an excellent diet, although growth will be slower. Such a feeding plan would particularly suit the slower-growing breeds.

In all cases there is little chance of overfeeding fattening ducks as the more they are given the more weight they put on. In the right conditions ducks grow at a fantastic rate. But where 'wet' feed is offered, ensure that a surplus is not left hanging around for it will soon go stale and unpalatable and attract unwanted guests like rats and mice.

Feeding geese

The degree of attention needed to goose feeding, for fattening or egg laying for breeding purposes, depends very much on the amount and quality of grazing available. In fact, the whole pattern of the geese industry is geared to the grass growing season. Food conversion of geese is such that it is not economic to feed them on grain, apart from in the initial brooding period. This is why the rearing season starts with hatching in March, with the goslings emerging from the brooders just in time for the first bite of grass. When growth slows up in the autumn, the writing is on the wall for the goose and most of them are killed for market. These are known as Michaelmas geese—killed by the feast of Michaelmas (September 29) and fattened entirely on grass. If they are taken closer to Christmas they need some additional grain.

Geese will thrive on short, green grass and when there is plenty available no other feed is really necessary; but they could starve on long coarse grass. They also eat weeds, like thistles and buttercups, and providing that too many geese are not put in too small a space, they can improve the quality of the land. If you overstock though, you'll end up with a quagmire.

If grass is in short supply the chances are there will be a green or common nearby that would benefit from a daily visit from your geese. Check first, however, to see that the common is not used by cattle. Cows will not use land that has been grazed by geese—they object to the goose's liberal output of droppings.

Starting the young

Since it is likely that you will find gosling starter pellets hard to come by, chick starter crumbs or pellets (without coccidiostat) are an adequate alternative. This feed should be made freely available at all times and sprinkled with grower-size insoluble grit. Although goslings will want to graze after a couple of days, hold them back for two to three weeks and then allow them access to short grass only. Make some flint grit available right from the start and through the growing period.

Where there is plenty of good grass about the amount of bought feed can be gradually reduced after the first couple of weeks, although some should always be available. Chick crumbs can be replaced by a home-produced mixture of biscuit meal and ground oats, with chopped lettuce, dandelions and green leaves, until it can be seen that the young goslings are getting enough grass. Wet mash or crumb feeds should be given as early as possible in the morning and again as late as possible at night.

Supplementary feeds can be stopped from about eight weeks if the grass is plentiful. If fast growth for fattening is wanted, give $2\frac{1}{4}$ oz each of dry or soaked mixed corn. Wheat, barley and oats is the cheapest combination. A small feed in the morning will encourage goslings into the range to eat grass, and a heavier feed of wet mash can be offered in the evenings. Mixed grain can be offered freely all day for rapid growth.

When the grass begins to die back in October, the birds can either be killed off and held in the freezer until wanted, or grown on, in which case supplementary feed will be needed in the form of mixed grain and/or mash. For a really special product, birds can be penned up three to four weeks before killing and fed on mixed grains, wet home made mash, boiled fish scraps (see duck fattening page 67) with some fresh cut green foods and grit. Skimmed milk added to the meal will give an improved finish. It is worth noting that if grazing is limited and the birds have to be confined to a run, you can do a lot to help by bringing in large amounts of cut greenstuffs and by adding a grain balancer ration. The only problem is the state of the run—it could become a liability if it is not given a chance to dry out properly so a system of alternating runs will make life a lot easier.

Breeding geese

Geese kept for laying can be reared in much the same way as those for fattening. From eight weeks onwards, if grass is plentiful little else needs to be fed apart from small amounts of either bought-in layers' mash or home-mixed mash. During the breeding season, or when egg laying starts, a handful of grain (2¼ oz) should be used to supplement the grass. During winter months, the birds may be fed on oats and leafy hay. The grain should be fed in the mornings and evenings and in quantities that can be cleared up in ten minutes.

Four weeks before hatching eggs are expected, which is usually early March, a special breeders' ration consisting of wheat, fish meal, maize, soya, grass meal, yeast and vitamin and mineral supplements is fed. Geese consume 9–10 oz of this each day and, assuming a goose-breeder ration or equivalent chicken-breeder ration (preferably in pellet form) can be obtained, egg production will be a fairly costly exercise. When spring arrives and the grass is growing and the geese are laying, the expensive breeder ration can be supplemented with a cheaper, home produced variety, combined with mixed corn.

After laying, breeding geese can be put out on range with nothing more to eat than grass in summer and autumn. But they will benefit from being put back on to a chicken layers' mash in December and breeders' pellets or mash in January.

At all times, remember the need for grit and a plentiful supply of drinking and head-bathing water.

Feeding turkeys

Just as with fattening ducks, there are a number of approaches to fattening turkeys, generally differing in the cost of feeding. As mentioned at the outset, the choice lies between maximum growth rate at high cost or slower growth rate at lower cost.

Whatever the choice, you cannot stint on the feeding of young poults. Get them off to a good start and the rest will fall into place. Day-olds must be launched on good-quality turkey-starter crumbs (or broiler crumbs if that is all you can get), which are high in protein and contain anti-blackhead and anti-coccidial preparations (see pages 83–84).

Young poults are notoriously reluctant to eat and drink in the first few days of life, primarily because they have bad eyesight and are nervous. For the first five days they can survive on food left over from the hatching egg, but if they have not eaten by then, death is the most likely result. So make things easy for them from the start. Put the crumbs on to half-dozen size, fibre egg cartons or cake tin lids so that they are easily accessible. The cartons are preferable to open trays since the poults are less likely to walk all over them, but they should be changed for new ones every three or four days. Make sure that water is available all day.

Frequent handling in the early days will confirm whether the poults are, in fact, eating. You can tell by the presence or absence of a bulging crop, below the 'throat' of the poult. If the crop feels empty, the first thing is to get some water down the bird's throat. This can be done with a small glass pipette with a rubber teat. Water can be put straight into the crop by pushing the tube gently down the throat.

Poults can be attracted to the feed by tapping the container with a finger. The noise and movement will arouse interest. Also, a stick or shiny pencil run through the trough will bring the crumbs to their notice.

Up to six weeks, the crumbs should be fed as they come from the bag and made freely available from the start in open troughs alongside the egg-tray feeders. From six weeks on, the expensive starter crumbs can be made to go further by the addition of moistened stale bread, confectionary waste and the like. Under this system a $\frac{1}{2}$ cwt bag of crumbs will last six turkeys about ten weeks.

The expensive way to fatten birds, under intensive conditions, is to go on to a turkey grower ration at about nine weeks with the choice of a finisher from, say, eighteen weeks onwards. If really heavy (26 lb) birds are being grown, this is the sort of programme the big boys follow.

A compromise from about eight weeks of age is to cut back severely on the amount of starter crumbs offered and feed instead whole barley or wheat, preferably soaked overnight in water, and to mix in with it grain balancer rations that can be bought from the corn merchant. These are concentrated rations containing extra proteins (22–24 per cent) and vitamins. At eight weeks of age the balancer can be fed straight, mixed in with starter crumbs. Gradually, from ten or eleven weeks, whole grain should be introduced until, by about sixteen weeks, balancer and grain are being fed in roughly equal proportions and starter crumbs are left out completely. By twenty weeks of age, grain can form two thirds of the diet.

Insoluble grit is required by turkeys, but should be restricted to a sprinkling on the food troughs in the early stages and fed sparingly until sixteen weeks when more grain is being fed.

Turkeys are also fond of green food and when allowed free grazing will consume up to 15 per cent of their food intake from succulent green crops, clover and lucerne leys. But any green food should be stopped within about two weeks of killing time.

As a rough guide to the amount of feed you will need, turkeys under ideal conditions and given the best feed will put on about 1 lb of body weight from every $2\frac{1}{2}$ to 3 lb of feed. Thus an 18 lb turkey (live) will consume in the region of 50 lb of feed. This conversion of feed into body weight is more efficient in younger birds and there is a point at which the conversion becomes so inefficient that there is no point in trying to put on any more weight. It is impossible for the smaller flock owner using cheaper feed to estimate when this will occur, but if large amounts of bread and scraps are being fed, the best guide is probably age.

Start killing at around twelve to thirteen weeks when your birds should weigh in the region of 12 lb live, and kill them off at intervals up to about twenty weeks, when they should be in the region of 15–17 lb. For this sort of régime you will be better off with medium-type turkeys (see page 58).

Under the heading 'Feeding fattening ducks' page 67, much was made of the use and value of household waste and cheaper feeding. Hard and fast rules do not exist, but the comments can be equally applied to turkeys.

Possibly for the first couple of crops, it would be wise to stick to a fairly standard ration until you have the hang of the job and can then start improvising with cheaper substitutes. The vital point to note is that there are plenty of substitutes for expensive compounded feeds, particularly potatoes and bread, which will provide an acceptable fattening ration but a slower than optimum growth rate.

6 Taking Care of the Investment
Predators, rodents and insects
Foxes and other predators
Foxes are a real threat to your farming enterprise

whether you are in the country or suburbia. You may not have seen many, or any, in your particular location but that is no guarantee that they are not about. The presence of ducks, geese or turkeys in your garden will soon attract them. A south Londoner, with three chickens and three ducks in his large garden, shot eight foxes in the space of a year and still some escaped his gun to eat his stock.

Unless you can provide sturdy battlements in the form of a 6 ft high wire fence around your garden, and bury the wire about a foot into the ground, birds running free will always be at risk. Assuming the house is sturdy enough to keep out predators, shutting them up at night will go a long way to ensuring their safety since foxes generally strike after dusk. The other alternative is to provide a covered run so that access from the outside will be virtually impossible.

In rural areas, stoats and weasels can present a similar hazard, particularly where very young stock is concerned. Cats, too, have been known to attack ducklings, but more often their presence in a duck run is merely a worry to the birds which can put them off lay, check their growth rate, or cause them to go lame.

The golden rule, unless the animals are being given a free run of the garden and made pets of, is to keep cats, dogs and children well clear.

Rodents

Although rats have been known to attack young stock and eat hatching eggs, the main concern will be to prevent a build-up of infestation. Like mice, they will be attracted by warmth and dark corners and, of course, feed. Even if your neighbour is prepared to put up with the noise of your 'farm', and the odd smell, he is well within his rights to call in the environmental health inspector if rats start appearing in his garden. Such a move could well stop your business, however much you apologise.

One way of dealing with an intrusion of rats is to set tunnel traps at strategic points around the duck house, preferably near freshly dug holes where they may be nesting. Poisoned baits can also be effective, but if it becomes obvious that a rat population is building up despite your efforts, seek the help and advice of the local authority health department or one of the pest control firms listed in 'Yellow Pages'.

Mice are easier to manage and are less likely to set up homes and families in duck or turkey houses than they are in chicken houses. However, they can be a nuisance and require prompt action. Start by making the houses mouseproof, ensuring that holes, even those as small as $\frac{1}{4}$ in. diameter, are blocked up.

The next step is to establish permanent baiting points to kill any invaders before they can multiply. Anti-coagulant poisons (like Warfarin) are of little use since poultry foods are rich in vitamin K which is the antidote. Two safe alternatives recommended by the Ministry of Agriculture are a 4 per cent alphachloralose mixture and Sorexa CR.

To prepare permanent alphachloralose baits, it is best to incorporate them in wax blocks as follows: mix 18 oz of whole wheat, 18 oz of porridge oats, 16 oz of pinhead oatmeal, 16 oz of dehusked canary seed, 3 oz of icing sugar, 5 fl. oz of corn oil and 4 oz of alphachloralose. This bait should then be mixed into 20 oz of melted paraffin wax (20 oz blocks of paraffin wax are available from some of the larger chemists). When set, the wax should be cut into 1 in. cubes which are then ready for use.

These paraffin bait blocks are particularly useful as they can be set in places that would be inaccessible for laying ordinary bait. They may also be put in suitable containers to provide permanent baiting points, which will keep the bait free of dust and palatable. An empty tin, with an entrance hole of about 1 in. square cut in it, is quite suitable simply nailed to woodwork in the corners, or weighted firmly beneath the houses or in the droppings pits with bricks.

It is a wise precaution to gather up and safely discard any leftover bait when treatment is over. Always keep fresh or discarded bait away from pets and other animals.

An effective alternative, although slightly more expensive, is Sorexa CR—a mixture of the poison Warfarin and calciferol (vitamin D_2). It will kill warfarin-resistant mice and although generally 100 per cent effective it can lead to poison-shyness, so a large number of baiting points may be required.

Possibly one of the best defences of all is a large measure of good housekeeping. Rats in particular harbour beneath pieces of timber, in nettles and in similar undisturbed spots, long before they take up residence in poultry houses. So everyone, including you and your stock, will benefit from a 'tidy surround' policy. Certainly, it will be one more safeguard against marauding rodents.

Insect pests

Fleas, lice and mites are all potential problems for ducks, geese and turkeys, but under favourable conditions of a good-sized, clean run they are unlikely to be a cause for concern. The birds are expert at preening themselves. If they are obviously suffering intense irritation, treat them with a proprietary dusting powder.

Diseases and health

Optimists say that ducks, geese and turkeys are such hardy creatures that there are no diseases worth considering. Certainly, once the birds have reached the point when they are off heat, nine times out of ten they should have a clear run. But things can go wrong, sometimes for no apparent reason and any deaths will not only upset you as a 'farmer' but also as an accountant.

Often, problems arise more from management failures than from actual disease organisms. With small numbers of birds, bought from reputable suppliers, you would be unlucky if the birds contracted infections. There are one or two possibles that will be discussed later in this chapter, but largely the onus is on you to manage and protect your charges to the best of your ability.

The golden rule is to ensure that the birds *always* have clean, fresh water available. In winter, de-frost the water troughs when they freeze over.

The same goes for feeding. Make sure the birds are getting enough but do not over-feed them. Do not leave stale food about, and ensure that very young stock get their share. Encourage those that are reluctant to feed. Make sure that grit is always available to aid digestion and limit the amount of whole grain feeding as it can cause upsets when available in large quantities. Get into the habit of handling the stock occasionally so that you can tell if the birds are in good, fleshy condition. The situation may arise where birds are eating well, but not thriving—many continue to feed even when sick.

A good guide to health is general condition—bright eyes, lustrous feathering and general 'bouncy' gait. The exception is with laying ducks. The best layers in the flock look rather poor, with particularly rough-looking feathers. Laying stock that look in 'tip-top' condition are almost certainly not doing their job.

General ailments

Chills. All young stock are susceptible to chills, so ensure that brooding accommodation is not draughty; at the same time, it shouldn't be stuffy. Pneumonia is a possibility in cold, wet weather, particularly if the birds are not fully feathered. Sufferers may not reveal their symptoms until the condition is well advanced, but if affected birds can be detected then some successful recoveries can be achieved by wrapping them up and placing them in a warm straw-lined box.

Sunstroke. Ducklings and goslings cannot stand long exposure to intense sun—they must have access to shelter or they suffer and may die from sunstroke. If there is no natural shade, provide them with an awning or give them access to the house.

Non-specific respiratory troubles. Respiratory infection is another hazard with stock under two weeks of age. Known as aspergillosis, it can kill small ducklings or poults swiftly and leave no trace. Laboured breathing is a symptom in older birds. The commonest cause is mouldy straw and special care should be given to the selection of bedding material.

Worms. Intestinal worms are a constant threat to all livestock allowed free access to the great outdoors. Many worms spend part of their lives in the variety of insects that may live on the stock and others are picked up in the run. They can infect stock when a healthy bird pecks at the droppings of a diseased bird.

One of the commonest worms is the gizzard worm in geese (see page 82) and ducks and turkeys are susceptible to a variety of species. Some of them can be tolerated in small numbers, but cause a severe check on growth and definite illness when present in force.

All-purpose worming powders are available to rid birds of the parasites, but a lot can be done in terms of management to cut the risks: provide a well-drained area and prevent the build-up of droppings in the shelter by moving the shelter frequently and by ensuring there is plenty of good quality litter on the floor.

Cramp is usually confined to young and adult ducks and geese. It occurs without warning and a bird is suddenly found lying on the ground, unable to use its legs and only able to move by flapping its wings. Twenty-four hours in a hay-lined box can often act as a cure.

Lameness is fairly common in geese, since these birds are fairly clumsy and often injure themselves. On occasions they may also be injured on catching, particularly if they are caught by one leg. But they usually recover.

Swollen foot, or *bumble foot*, can affect all stock on a free range, particularly on dry, hard ground. It may be caused by a piece of cinder or stone becoming embedded in the foot. A swelling is created on the foot and can spread up the leg causing lameness. If found in time, tincture of iodine should clear the problem.

If this fails, the swelling can be incised with a sharp, sterile knife, making a small cut and squeezing or scooping out the contents. Clean the wound thoroughly with warm water in which a few crystals of potassium permanganate have been dissolved, then bandage the foot.

Slipped wing can easily occur in geese and is recognised by the flight feathers protruding at right-angles to the body. It does not seem particularly distressing to the birds, and although there is no cure, the feathers can be chopped back. Young goslings are also prone to *drooping wing* which is believed to be the result of a fright. The wing may drag on the ground, but the birds usually recover.

Distended crop. Occasionally in ducks, geese or turkeys, the front of the lower neck region becomes swollen. If the swelling is hard to the touch it is probably a case of crop-binding resulting from bits of straw, hay or other stringy material clogging the outlet of the crop. Feed the bird warm water with a teaspoon, then hold the head downwards and gently squeeze out the contents of the crop. Follow this with a teaspoonful of olive oil. Repeat the treatment the next day if necessary.

In sour crop, when the crop contents are soft, they can be expelled right away without added water. Warm water should be given later with a pinch of bicarbonate.

Feather pecking. This condition, which can lead to cannibalism in chickens, is rare in ducks, turkeys and geese, particularly if they have plenty of room to move. The best way to avoid it is to keep the birds occupied— have some greens hanging up for them to peck at as a distraction, while the odd tin can in the run will give turkeys hours of 'pecking pleasure'. If, despite your efforts at prevention, the birds lose feathers and blood starts to appear, paint the affected parts with a black treacle called Stockholm tar, obtained from pet shops. Alternatively, mix your own anti-peck cream from 4 oz of petroleum jelly, $\frac{1}{4}$ oz of aloes and $\frac{1}{4}$ oz of carmine.

Duck ailments

Prolapse. Very occasionally, laying ducks suffer from prolapse, a condition in which a portion of the internal organs protrudes through the vent. It may occur when the birds are endeavouring to lay or just after laying. There is a slim chance that the bird can be cured, but the condition often repeats itself and the only answer is to destroy the duck.

To treat the condition, first feel the duck to ensure that there is no egg inside. If there is an egg, handle the bird very gently. Grease the protruding organs with vaseline and try to work the egg slowly along the passage by gentle manipulation. Once this has been done, try to replace the organs. Wash the hands thoroughly in disinfectant and gently try to replace the exposed parts. Do not use force. If you are successful, isolate the bird for a few days and give her light meals with no grain and plenty of warmth. If not, there's really only one thing you can do—prepare the bird for the table.

Duck Virus Hepatitis. It is worth knowing something about this highly infectious disease, even though there is little that can be done about it. It is caused by a virus that affects young ducklings and can kill within two or three days. Affected birds stop eating, although they drink well, and their movements become rather sluggish. In a day or two they fall on their sides with heads drawn back and paddle spasmodically with their feet. While some birds die, others make a complete recovery.

There are preventative vaccines available, but veterinary advice is needed on their use.

Pasteurellosis. This again is a relatively rare disease. It is caused by a microscopic organism that strikes birds over four weeks of age. Symptoms are loss of appetite, thirst, high temperature and a sort of diarrhoea coloured with green bile. 'Knee' joints may also develop swellings and a few birds may die.

These are only guidelines to the symptoms since positive diagnosis is a job for a laboratory. In truth, there is little point in going to any great lengths to have it diagnosed unless all birds are affected—and even then you will have to consider the cost carefully.

Geese ailments

Gizzard worms. Geese can pick up small parasitic worms from the land, particularly if they have been using it for a number of years and their grazing area has become infested.

The worms are reddish brown, about $\frac{1}{2}$ to 1 in. long, and bury themselves in the lining of the gizzard. Large numbers can cause a great deal of discomfort. Young birds are particularly susceptible and, once infected, start to lose weight or grow less well than the rest, becoming thin and weak. They may also walk with a stagger.

A commonly recommended treatment is to 'force

feed' the goslings with a 1 cc or 2 cc capsule of carbon tetrachloride, repeating the dose seven days later. During, and for a few days after, the treatment they should be fed well on a good layers' mash mixed with milk. Do not exceed the recommended dose since carbon tetrachloride is poisonous.

Turkey ailments

Blackhead. This is caused by a microscopic parasite that gets into the intestines. Blackhead was once the scourge of the turkey industry until a drug was discovered that could be included in the feed. Although commercial producers use feed with an anti-blackhead drug mixed in, there is no doubt that the disease is still very common and must be considered a major risk with a backyard flock.

Where proper turkey rations are being fed, you can ask for an anti-blackhead drug to be included. Make sure that it is written on the bag and feed it throughout the birds' life, with the exception of the last week or so before you intend killing.

One of the drawbacks of 'diluting' the ration with scraps and the like, or using broiler feed for example, is that birds will only get a fraction of the intended dose of drug at each feed, or none at all. This is a risk that has to be considered.

There are soluble drugs available that can be added to the water or to the home-mixed feed during preparation.

The first signs are a decline in weight and loss of appetite. Affected poults grow weak and drowsy and stand about with ruffled feathers, drooping heads, wings and tails. Sulphur-yellow droppings are also a characteristic symptom. Death can be quite sudden.

The disease organism can be passed on to turkeys in the eggs of a small worm that infects chickens. The golden rule, therefore, in the control of blackhead, is not to run turkeys on land that has been used for chickens. Similarly, alternate the run if possible so that turkeys are not reared on the same ground two years in succession. Do not rear young birds near or on ground used by older birds. Since earthworms can harbour the disease, treat the run with insecticide to reduce worm numbers at the first sign of blackhead.

Coccidiosis. This is a disease which affects the intestines of turkeys and, although not a killer, has similar symptoms to blackhead. Birds stand around, look ruffled and lose weight in the same way, but the droppings are more likely to be watery, grey to brown in colour, often with green bile stains.

Drugs to control the disease, called coccidiostats, are available and can be added to the feed at low levels according to the manufacturer's instructions. If ready-made turkey feeds are being used, it is worth asking for a coccidiostat to be included along with an anti-blackhead drug. The alternative approach, especially with cheaper feeding, is to have a coccidiostat on hand so that it can be given in the drinking water as soon as the condition is suspected.

Newcastle disease. Most people have heard of this condition under its common name—*Fowl Pest.* It is a disease that has to be watched for and, if suspected, reported to the nearest Ministry of Agriculture veterinarian. This is one poultry disease that you are required by law to report to the authorities.

It is caused by a virus that attacks the nervous system. Affected turkeys show signs of gasping, coughing, rattling of the windpipe, loss of appetite, huddling and nervous symptoms, like paralysis of wings and legs, and apparent dizziness, the head often being bent either forward or back or rotating.

Fortunately, there is an excellent preventative vaccine available to hold the disease in check. It is widely used in the commercial world but one drawback for the small-scale farmer is that the smallest quantity available will treat 250 birds. The vaccine has to be made up and, once mixed, will not keep for more than a couple of hours in a cool place. You have two choices—throw the surplus away or co-operate with a neighbouring poultry keeper and use part of the supply on his birds as well.

The Newcastle disease vaccine used for turkeys is Hitchner B1 which is given at around two or three weeks of age. If you buy growers, ensure that they have been vaccinated already.

To apply the vaccine, follow the makers' instructions. Briefly, the vaccines are given in the drinking water after a thirst has been created by leaving the drinkers empty for two or three hours. Mix the vaccine pack in about a gallon of clean water in a clean, plastic bucket, which contains no traces of detergent. About half a pint of the mix should be added to the drinker for six birds. If this is consumed rapidly, add a little more.

Preventing the problems

Way before diagnosis and cure comes the small matter of prevention of ill health. A great deal can be done by the observant and keen stockman.

Most of it is commonsense: like watching the birds carefully for any signs of discomfort, and making sure the run does not become too wet and too muddy. If possible, use the run in rotation or alternatively dig it over occasionally and treat with lime and insecticide. Better still, cover the surface with 6 in. or more of cinders or shale to give a good, free-draining surface.

Keep the water and feed as fresh as possible and ensure that birds are not short on feed and are never without water. Keep the utensils clean: wash them every day and sterilise them weekly with boiling water.

Never use mouldy hay or straw as bedding or litter material. In bitterly cold weather, do something about 'wrapping up' the simpler type houses with sacks and old rugs.

As far as the bird accommodation goes, make sure an earth floor is kept as dry as possible with plenty of clean, dry straw for bedding. Between crops of birds, scrub the house down, first with water and washing soda and then with disinfectant. Finally, give the outside a coat of creosote once a year.

All these things are plain commonsense—but then so is stockmanship.

7 The Duck Egg Harvest

If the management is right, a healthy duck will lay between 250–320 eggs in its first laying year. If the moult can be controlled to last five or six weeks, it will go on to lay 25 per cent fewer eggs in the second year than the first. There will be a similar reduction in the third year.

Twelve ducks will lay at a good rate for the three years without the need for replacement. In the first year, you can expect virtually an egg a day from each bird—bear this calculation in mind when planning your enterprise. Surplus eggs can be sold to friends, neighbours and local shops, but research your market first.

Collection and storage
Duck eggs need more careful handling than chickens'

eggs, primarily because the pores in the shell are larger, giving bacteria a better chance to enter as the egg cools after laying. Research work in the USA has shown that, given ideal conditions in terms of egg quality (fresh, clean nest litter and frequent collection), a duck will lay an egg with a better shelf life than a chicken's egg. In practice, however, duck eggs should be eaten within two weeks of being laid. If you are selling part of the output, do not take money for an egg that is more than one week old.

To know the age of your eggs, it follows that they must be collected every morning immediately the ducks are let out for the day. Most of the eggs will have been laid by about 9.00 a.m. but, to be on the safe side, check the run each day in case one or two birds have a favourite laying spot outside. Unless you know for certain when outside eggs were laid, destroy them. If you are holding eggs, note the day of collection, place them on Keyes trays and store in a cool place.

Nest boxes will help keep the eggs clean if the birds can be persuaded to use them. Check the nest box litter daily and change it whenever it's dirty or damp. If birds lay their eggs on the floor, ensure that the litter material is clean and dry. Small numbers of eggs are best collected in a bowl or fibre egg pack; for larger numbers use a thirty-egg capacity Keyes tray or a rigid basket with some cotton-wool at the bottom.

When collecting the eggs remember that ducks like a calm, ordered existence. Surprises and shocks send them into a panic in which eggs can get broken.

On collection, keep dirty eggs separate from the rest and clean them immediately with a nail brush under lukewarm, running water. In fact, give all the eggs a rinse under the tap immediately after collection and then dry them with a clean cloth. The job will only take a few minutes and is well worth the effort.

Always store eggs with the pointed end downwards

on racks or Keyes trays. Do not hold on to cracked eggs for too long as these are likely to 'go off' sooner. Store all eggs away from onions, fish, cheese or any other strong-smelling foods. If they are stored in a fridge do not take them out until they are wanted or the internal quality will deteriorate rapidly.

Badly cracked eggs can be lightly beaten with half a teaspoonful of salt or sugar per egg and placed in a fridge in a covered container. Separated yolks can be similarly treated with half a teaspoonful of salt or sugar to two yolks; the salt or sugar prevents the yolks becoming sticky. Egg whites can be stored on their own. Label the container as to the number of eggs involved and whether sugar or salt has been added. Eggs can be stored for up to ten days if placed in a fridge in this way.

Treated in this fashion, eggs will last ten months if kept in a freezer. But use top-quality eggs that have not been exposed to dirt. If you have had to wash eggs to get them clean, do not use them for freezing.

Another idea is to freeze lightly beaten eggs in an ice-cube tray. Once frozen, the cubes can be stored in a polythene bag, each cube representing the equivalent of one standard chicken egg.

Separated yolks and whites can also be frozen in the ice-cube tray. For cooking purposes, one tablespoonful of thawed yolk equals one standard egg yolk, but it takes two tablespoonfuls of thawed egg white to equal one standard egg white. Don't forget the labels.

Water-glass storage, using sodium silicate, is still popular when a freezer is not available. This low cost chemical is available from chemists and hardware stores, complete with mixing instructions. It is dissolved in water in an earthenware or enamel vessel and clean, sound eggs added. The water glass seals the shell, preventing loss of moisture from inside and keeping out bacteria. Water-glass stored eggs are best used for cooking once they have been washed under running water.

Egg quality and faults

Soft shells. These are common at the start of lay when the birds are settling into their routine. They may also occur towards the end of a laying year, prior to a moult. If the problem persists, check that the ducks are receiving calcium in the form of oyster shell grit.

Cracks. Cracked eggs are usually the result of some management failure, such as mishandling during collection, insufficient soft material in the nest box, or birds being disturbed in lay. Ducks rarely lay cracked eggs.

Mis-shapes. Flat-sided, bulging, wrinkled or rough-shelled eggs are all natural phenomena. As such there is little you can do or indeed need do about the problem, except eat the eggs—they are perfectly edible.

Dirt. Don't be alarmed if egg shells carry smears of blood; this is normal with young layers. If shells are particularly dirty check the litter in the nest box or on the floor.

Internal quality

The guide to internal quality is the condition of the egg when it is broken out for frying or whisking.

Runny whites. If the white runs all over the pan, the egg is either old or has been stored at too high a temperature. It could be related to improper feeding but this is unlikely. Thinning of the white is one of the most common characteristics of the ageing process in eggs. In a fresh egg the yolk stands proud of the white which forms a tight outside circle. Later in life the white spreads and the yolk flattens out (see Fig. 18).

Pale yolks. An unlikely problem caused by lack of grass meal in the diet. Give the birds some grass cuttings or other green stuffs.

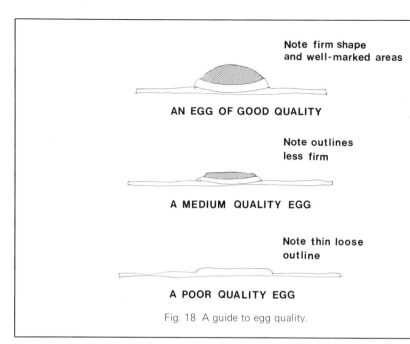

Note firm shape
and well-marked areas

AN EGG OF GOOD QUALITY

Note outlines
less firm

A MEDIUM QUALITY EGG

Note thin loose
outline

A POOR QUALITY EGG

Fig. 18 A guide to egg quality.

Blood or meat spots. The egg contents may be marred by small blood or meat spots either in the yolk or white or more commonly sticking to the edge of the yolk. They are the result of minor haemorrhages as the egg is formed or by small pieces of tissue fragmenting. They are neither harmful to the egg eater or the duck —just unsightly.

Testing for quality

If you are selling any of the eggs, you need to set quality standards and provide only the best. This means that the cracked ones will be eaten at home along with any older eggs. A quick test on age is to hold the egg up to a strong light and check on the size of the air space compared with a known fresh egg. There will be a marked difference if the egg is a week or ten days old. The contents will have shrunk and the space enlarged.

While the egg is in front of the light, check for meat and blood spots, and fine hair cracks in the shell that escape the naked eye. Technically, these are second quality goods and bad for your reputation as a purveyor of top quality eggs to the gentry.

A torch beam and a darkened room can be used to check internal quality. Hold the egg between thumb and first finger and gently rotate it in front of the beam. You will soon pick out any blood spots and cracks. Candles were once used for this operation, hence the term 'candling'.

A more sophisticated and permanent device is a

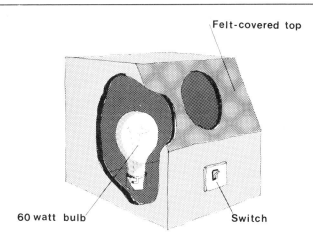

Fig. 19 A home-made egg candler. All that is needed is a 60 watt bulb, a bulb holder connected to the electricity supply, a switch and some spare pieces of plywood. Make the candler about 10 in. × 10 in. with a 1 in. hole on the angled side and a felt covering around the hole. Switch on and take a closer look inside the egg by holding it broad-end uppermost at a slight angle to the aperture, and twirl it on its axis.

home-made candler like that shown in Fig. 19. The source of light is a 60-watt, clear glass bulb set in a 10 in. × 10 in. box with one sloping face carrying a 1 in. diameter hole. The felt cover reduces leakage of light round the outline of the egg.

Hold the egg broad end uppermost at a slight angle to the aperture of the lamp and twirl it on its axis. In a fresh egg the air space is $\frac{1}{4}$ in. deep and the yolk is set firmly in the centre when the egg is rotated. In older eggs the air space is larger and the yolk moves appreciably within the white.

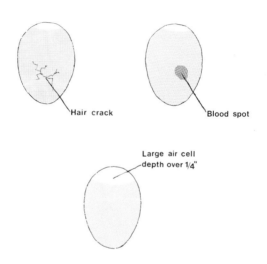

Fig. 20 Egg faults—these are some of the faults that will show up if you use an egg candler. None of them is harmful and the eggs will still taste good, but anyone selling quality eggs would have to remove these from the batch. The hair crack could well result from rough handling; the blood spot will occur from time to time; and the enlarged air cell indicates that the egg is not particularly fresh.

3 Making Ready for the Table

Killing and preparation of your birds for the oven or the freezer is the one area of garden farming where you may reasonably consider that outside help is needed. The technique of killing geese and turkeys, and ducks to a lesser extent, only comes with practice. Ideally, therefore, it will be easier if you can take advantage of someone else's experience on the first few occasions.

When you kill will depend to a large extent on requirements, particularly if you intend eating a bird at a specific time like Christmas or Easter. More correctly you should kill birds when they are ready, but only experience will tell you when the ideal time has arrived.

Killing times

The secret with ducks is to have them ready from seven to eight weeks of age when they will weigh in the region of 6–8 lb live, giving a $4\frac{1}{2}$–6 lb bird for the table. Feathering is one of the ways of checking if the birds are ready. If they are completely feathered and the feathers look hard and bright, they are ready. Run your finger up the breast of the bird. If you can feel stubs, it is not ready. Nor is it ready if there are still patches of stubby down on the breast and vent. Flight feathers should be $3\frac{1}{2}$–4 in. long at maturity.

Stubs make plucking difficult. The duck has its first set of feathers and down at eight or nine weeks of age. It then sheds its body feathers but by ten or eleven weeks new ones are growing and the emerging feather stubs are the problem. To make life bearable for the plucker it is necessary to wait until the birds are around sixteen weeks old.

The rearing period of seven to eight weeks for ducks assumes that the birds are on full rations. If 'low level' feeding is practised, feather development will be delayed a week or two and killing should depend not so much on the calendar, but rather the feel and weight of the birds.

Laying ducks, of course, can be eaten as well as meat birds and should be plucked and prepared in the same way. Probably the only difference comes in the cooking. Older, end-of-term layers will be tougher and should be casseroled for best results.

Geese also have stub problems, but in their case the killing date will depend more on the plumpness and colour of the breast. Target weight should be about 12 lb plus at around twenty-two to twenty-four weeks. Some breeds fed on good quality rations will reach 18–20 lb in about thirty weeks. But it will take about 4 lb of feed to get 1 lb of goose flesh, which is not really an economic rate of conversion.

Turkeys, fortunately, are better converters but their rate will also be slower if a cheaper ration is fed. Even so it is reasonable to expect a twenty-week old bird to weigh about 16–17 lb liveweight and about 10 lb oven-ready. You can, in fact, start picking birds out at around twelve to fourteen weeks when they will weigh about 9 lb liveweight and $6\frac{1}{2}$ lb oven-ready.

Method of killing

Unlike condemned men, ducks, turkeys and geese should not be allowed a hearty breakfast before they are to be killed: they should be starved for the final twelve hours before execution. This is not only as an economy measure—they are not going to survive long enough to convert the feed into meat—but to enable the birds' last meal to clear the gut before the end arrives.

The best and most practicable way to kill a duck,

goose or turkey is by dislocating its neck. If done skilfully, death is instantaneous.

Ducks are the easiest to kill because they are smaller and easier to handle. Hold the duck by its feet in your left hand with the bird hanging head downwards (see Fig. 21). Take the head in the right hand so that the palm is on top of the bird's head and the thumb presses against the underside of the neck just below the head. Give the head a sudden jerk with the right hand, turning the thumb down and inwards. Dislocation is indicated by easing of the pressure. Hold the head there for a moment, then give a second controlled pull to extend the neck by an inch or two. This second pull will create a 2 in. cavity in the neck which will take the blood from arteries which have been severed at the same time. If you have not done the job properly, the neck and head will begin to swell and the second tug will have to be repeated, but with more conviction.

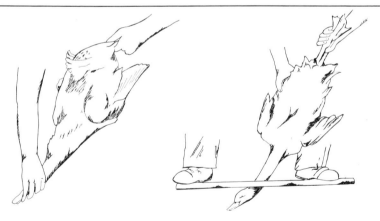

Fig. 21 *Left:* a duck's neck can be dislocated by holding the feet in one hand and pulling the head down with the other. *Right:* the same principle can be used for geese and turkeys, except that the bird's head is placed under a broom handle which can be held on the floor by the feet as shown.

As soon as the neck is dislocated, the body and wings will start to flutter violently. Although alarming to the novice, these movements are insensitive nervous reactions to the breaking of the spinal cord. Hold on to the legs or hang the bird by the feet from a convenient point until the spasm passes. Either way, the bird should remain head downwards for about three minutes to allow the blood to drain into the neck cavity.

The necks of turkeys and geese (ducks too, if it is easier) can de dislocated by the 'ground bar' method. It involves a pole, such as a broom handle, placed on the floor across the neck of the bird just behind the head, with the top of the head upwards and the breast facing you (see Fig. 21). Stand astride the bird and place both feet on the handle. Hold the bird by the feet and pull firmly upwards for a rapid, no-nonsense dislocation.

Again, it is vital that there should be complete severence, otherwise the body will not drain properly. It is possible to bleed the birds by making a slit with a sharp knife in the region of the dislocation. In view of the size and strength of geese wings, it is essential that before killing they are crossed twice across the back of the bird to hold them in place.

Out with the feathers

Plucking should be started as soon as possible after killing—feathers leave a warm carcase more easily than a cold one. It is a messy job, so wear old clothes and have an old bin ready to take the feathers. Ideally, do the job in a garage or a shed so that feathers are not blown all over the place, and arrange it so that the bird can be hung by its feet from a beam at a convenient height.

Normally, single birds are plucked dry but the process can be simplified by first immersing the birds in hot water. Wet plucking is certainly the rule if a number of birds have been killed at once. It also makes the job much cleaner since the wet feathers do not blow about.

The secret of good plucking and a good finish is to immerse the bird in water just off the boil. Tie some string around the bird's leg, plunge the bird in and hold it completely immersed with the aid of a stick, while you count to fourteen. Try pulling a few feathers to see if they come away easily.

Generally speaking, the approach to plucking will be the same for ducks, geese and turkeys, although ducks have their own peculiarity in that the down can be particularly difficult to remove. They are by far the most tedious of the three birds to prepare, although the job will become easier with practice.

Tackle the wing feathers first. (If you are dealing with older ducks or geese, leave the flight feathers because, during the final dressing process, you will have to cut off the last joint of the wing.) Next remove the tail feathers and then start to pluck around the vent and the legs. This will give time for the fat around the breast to solidify, giving more strength to the skin, which has a tendency to tear.

Beginners will find it easier to do the breast of a turkey before the legs and vent since the feathers come out much easier when the body is still warm. To avoid tearing, hold the skin with one hand, and with the other pull out small groups of feathers in the direction in which they lie on the body. Pluck from the neck upwards to the vent in a regular order. This should be repeated over the whole carcase except the head. Be particularly careful when plucking the tender areas on either side of the breast and thighs and in front of the parson's nose and around the crop.

Next comes the 'stubbing' where short feather stubs are removed with a blunt knife and thumb which act like tweezers. There may still be a mass of down and short feathers left on the body, particularly with ducks. The professionals cope with these difficult areas by dipping the carcase in hot paraffin wax for five seconds, then into cold water to harden off the wax. Stripping off the hardened wax removes the fine down and feathers at the same time. The wax can be reclaimed.

Obviously, where only a few birds are involved waxing is not economical or necessary. A satisfactory finish can be achieved by singeing, which works particularly well on turkeys and geese. The carcase is held over a methylated spirit flame which burns off the feathers, but does not taint the flesh. Another method for ducks is to cover the rough plucked carcase with a wet cloth and then smooth it over with a warm iron. Peel back the dried cloth and away will come the offending feathers.

Finally, squeeze the vent to get rid of any excreta and wash the feet.

As soon as the birds are finished they should be hung in a cool airy place (Fig. 22). How long you leave them before gutting (eviscerating, in the trade) is, within limits, up to you. Held at a temperature of no more than 40°F they will certainly keep up to ten days without any deterioration. It is worth holding them for a few days at least for it adds a flavour to the flesh. None of the frozen poultry available in shops today has been 'hung' for any length of time and the relative lack of flavour compared with hung birds is very marked.

To some extent, the decision as to whether you hold them or not will depend on your immediate plans. If the turkey is destined for the Sunday or Christmas table, the killing and dressing programme should be planned to include five to seven days hanging. If, on the other hand, you intend to joint the bird and keep it

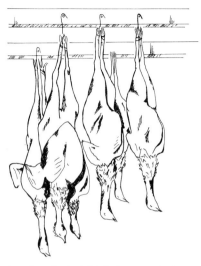

Fig. 22 Hanging the plucked birds in a cool airy place will definitely improve their flavour.

in a freezer (see page 101) then it is really advisable to prepare it soon after plucking and freeze it within a few hours. The same arguments also apply to ducks and geese, although short of cutting them in half there is not much scope for jointing the carcases.

Gutting

This is the process of removing the innards and can be quite a messy business. Equip yourself with a decent-sized chopping board, a sharp knife and some string. Knives, hands and working surfaces must be cleaned thoroughly before and after the operation; they mustn't come in contact with other foods or food preparation areas until they have been scrubbed clean.

First remove the sinews in the legs. (The only exceptions are young ducks killed around eight weeks of age when this will not be necessary.) This is done by cutting round the legs, about an inch below the hock. Break the bones at the point of the cut and when the feet are pulled away from the leg, the sinews will follow. The tissues may need to be loosened with a skewer to help the process. With older ducks and geese, cut through the joint between the two long wing bones and discard the end joint. In turkeys, there is some meat on this joint so leave it intact.

Evisceration of ducks and geese is a somewhat different process to that of turkeys. Place the duck or goose breast downwards with the head hanging over the edge of the table. Make a 3–4 in. long cut in the centre of the neck skin. With the left hand holding the neck, sever this where it joins the shoulder, leaving a flap of skin long enough to be turned back over the shoulder. With the neck itself out of the way, insert a finger into the neck cavity and work the crop round until enough is exposed to make a cut low down in the neck cavity; the crop can be cut away and the wind pipe removed at the same time.

Now turn the bird on to its back and insert the second finger of the right hand into the neck cavity to loosen the lungs, which are embedded in the ribs near the backbone. This will ease their subsequent removal, but do not take them out at this stage. Instead, stand the carcase on its shoulders and make a cut between the vent and the 'parson's nose'; this will expose the intestines. Insert a finger round the underside of the vent, which will help define its outline and enable it to be completely cut away from the body but leaving the intestines still attached.

With the bird again on its back, and the tail towards you, enlarge the initial cut in the abdomen by slitting a little way towards the breast. It is likely that substantial quantities of fat will be revealed. These should be removed before the hand is put right inside the body.

Grasp the gizzard and give a slow sustained pull to bring out the intestines in one piece plus the heart, liver and lungs.

The gizzard, heart, liver and neck can be kept for cooking after surplus blood and fat has been cleaned off. The gizzard also has to be skinned by splitting it lengthwise with a knife and removing the inner horny skin that holds food material. A duck's gizzard is particularly tough, but if it is boiled for a few seconds after splitting, the lining should come away quite easily. Finally detach the small, green gall bladder from the liver, taking care not to puncture it since the contents will be very bitter.

There are two major differences in preparing a turkey. Firstly, removal of the neck can be difficult as it is so tough. Make a cut between two vertebrae, all round the neck, and then bend the neck backwards to break it off—preferably over a table edge.

Secondly, the lungs, windpipe and crop cannot be loosened from the neck end. All the loosening and removal must be done from the back, which means cutting a larger hole around the vent. At the front end, stags, in particular, carry what amounts to blubber, which should be shaved off with a sharp knife.

Jointing and trussing

The bird is now ready to be prepared for the oven. Washing the carcase at this stage is not normally necessary unless the intestines or gall bladder have been punctured. Just give the carcase a wipe over with a damp cloth—inside and out.

The carcase can be either tied up, which will make it more presentable for cooking and freezing, or cut into portions.

Turkeys take up far less space and offer more scope in the way of mid-week meals if the bird is jointed. You can choose your own approach to this, but with a sharp knife it is easy to cut away thighs and drumsticks, wings and large fillets of breast meat, all of which can be popped into plastic bags, labelled and frozen.

Ducks, and to some extent geese, can be either cut in half for storage, tied up, or put into tight-fitting plastic bags which will help keep them in shape.

Rather than go to the lengths of using a trussing needle and tying a carcase professionally, you could follow this simpler method. With the bird on its back, take a 20–24 in. piece of string. Place it under the back of the bird, bringing it up under the sides of the body and the wings. Holding both ends of the string, bring them over and then under the legs, forcing them as far forward as possible. Now take the strings back to the front of the bird, over the wings and as close as possible to the body. Turn the bird over on to its breast and tie the strings tightly across the back.

The wings should be tied under the bird with another piece of string. Finally, tie the two legs together at the knuckle joints, tie down under the parson's nose and tuck the neck skin flap under the wings.

While the above approach will certainly suffice for a duck or goose, you may prefer to take a more professional approach for a turkey. The method is shown in Fig. 23 and you will need a 10 in. trussing needle and about 2 ft of string.

Because of their different frame and fleshing, ducks and geese require a different approach. With the skin flap tucked under the body, push the legs forward and press them down on the table while passing the needle through the top of the thigh, through the body and through the other thigh and across the back. Tie the two ends tightly.

Re-thread the needle and pass it through the skin just below the end of the breast bone, and then cross the string under the legs and tie firmly. Make an incision in the skin below this and push the parson's

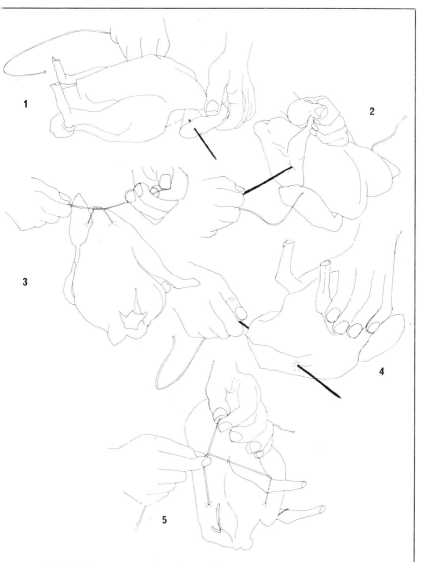

Fig. 23 Five easy steps in tying a turkey.
1. After threading the trussing needle, fold the wings over the back and pass the needle through them at the points shown.
2. Press the legs downwards and pass the needle through the body from thigh to thigh.
3. Bring the two ends of the string over the back and tie, making sure that the flap of skin is covering the neck hole.
4. Pass the needle through the skin under the pelvic bones as shown.
5. Take the string round one hock joint, through the skin at the tip of the breast bone, over the other hock joint and tie.

nose through it. This will come up through the hole made for removal of the intestines. The bird is now ready for the oven.

And so to the table!

The best way to cook ducks and geese whole is to prick the skin all over with a fork, sprinkle with salt and place in a hot oven. Cooking will take about $1-1\frac{1}{2}$ hours for a duck and 2 hours for a goose. Every half an hour or so the fat should be poured off and for the last ten minutes the oven should be turned up to brown the skin.

Turkey portions can be fried in batter or breadcrumbs, roasted or casseroled. The stripped carcase is an excellent source of stock for soups. The British Turkey Federation will be able to supply detailed recipes for cooking the portions if you run out of ideas. Their address is given in Appendix A.

As for that roast turkey Christmas dinner, you don't need us to tell you how to do that!

Bon appetit!

Appendix A

Useful addresses

Help and Advice

Animal Medics Ltd,
Regency House,
Greenacres Road
Oldham, Lancs.
Tel: 061–652 1307
(Stockists and suppliers of
poultry medicines, vaccines,
disinfectants and equipment.)

British Goose Producers
Association,
High Holborn House,
52–54 High Holborn,
London, WC1V 6SE
Tel: 01–242 4683

British Turkey Federation,
High Holborn House,
52–54 High Holborn,
London, WC1V 6SE
Tel: 01–242 4683

British Waterfowl Association,
6 Caldicott Place,
Over Winsford,
Cheshire, CW7 1LW
Tel: Winsford (0606) 594150
(An association of enthusiasts
interested in keeping, breeding
and conserving all types of
waterfowl. Services include
magazines, directory of
breeders, open days, bookshop
and advice.)

Duck Producers Association,
High Holborn House,
52–54 High Holborn,

London, WC1V 6SE
Tel: 01–242 4683

Eastern Counties Farmers Ltd,
PO Box 34, 86 Princes Street,
Ipswich,
Suffolk, IP11 RU
Tel: Ipswich 217070
(Suppliers of animal health
products in East Anglia.
Independent post mortem
service.)

Poultry Club of Great Britain,
Cliveden, Sandy Bank,
Chipping, Preston,
Lancs, PR3 2GA
Tel: Chipping (09956) 423

Salisbury Laboratory Ltd,
32 Grand Close, Downton,
Salisbury, Wilts.
Tel: Downton 22196

*Ministry of Agriculture
regional advisory offices –
advice, for which there
will be a fee, on all aspects
of poultry keeping.*

Northern Region:
Block 2, Government
Buildings,
Otley Road,
Lawnswood,
Leeds, LS16 5PY
Tel: 0532–674411

Midlands and Western Region:
Woodthorne,
Wolverhampton,
West Midlands, WV6 8TQ
Tel: 0902–754190

Eastern Region
Block B,
Government Buildings,
Brooklands Avenue,
Cambridge, CB2 2DR
Tel: 0223–358911

South Eastern Region:
Block A, Government Offices,
Coley Park, Reading,
Berks, RG1 6DT
Tel: 0734–581222

South Western Region:
Block 3, Government
Buildings,
Burghill Road,
Westbury-on-Trym, Bristol,
Avon, BS10 6NJ
Tel: 0272–500000

Wales:
66 Tyglas Road,
Llanishen, Cardiff
Tel: 0222–757971

Ministry of Agriculture veterinary investigation centres will carry out post mortem examinations at a charge of £8.50 for each bird or £13 for a batch of two or three. Regional offices or a telephone directory will provide the address of your local VI centre.

Dept. of Agriculture for Scotland, Poultry Advisers

West of Scotland Agricultural
College,
Poultry Dept.,
Auchincruive,
Ayr, KA6 5HW
Tel: 0292–520 331

East of Scotland College of
Agriculture,
Bush Estate,
Penicuik
Tel: 031–445 5353

North of Scotland College of
Agriculture,
Craibstone Farm,
Bucksburn, AB2 9TQ
Tel: 022–471 2677

Veterinary service
The veterinary surgeons listed below, with telephone numbers, provide a comprehensive farm, laboratory and post mortem diagnostic service. A comprehensive advisory service covers health, productivity, husbandry and environmental problems.

R. W. Blowey, Gloucester.
0452 24961
I. R. D. Cameron, Cheshire.
0829 40639
P. B. Clarke, Uckfield.
0825 4268

K. Gooderham,
Cambridgeshire. 0480 62816
G. Grant, York. 0904 643997
C. Harding, Lincolnshire.
06582 3276

H. Hellig, Colchester.
0206 572410
N. E. Horrox, Yorkshire.
0262 88692
P. W. Laing, Herefordshire.
0568 3232
J. O'Brien, Mildenhall. 717615
D. G. Parsons, Wiltshire.
0225 709184

D. Shingleton, Devon.
0548 830552
J. C. Stuart, Norwich.
0603 629046
O. Swarbrick, Arundel.
024368 2300
A. S. Wallis, Blackpool.
0253 899369

Stock

A county by county guide in England and Wales to the main suppliers of day-old and growing ducklings, goslings and turkey poults. There is a special section for Scotland.

Avon
G. C. Hope,
80 Montreal Avenue, Horfield,
Bristol, BS7 0NO
Geese: Toulouse

A. J. Newton,
Silvervale, 131 North Road,
Stoke Gifford,
Bristol, BS12 6PE
Tel: Bristol 792169
Ducks: Welsh Harlequin,
Khaki Campbell, Silver
Appleyard

Bedfordshire
B. E. R. Burman,
New Inn Farm,
Silsoe, Bedford
Tel: Silsoe 60868
Geese: Embden
Ducks: Aylesbury, Black East
Indian, Indian Runner,
Muscovey

Miss B. M. Cooper,
Chestnut Farm, 25 High

Street, Henlow, SG16 6BS
Tel: 0462 812280
Ducks: Welsh Harlequin

Mrs K. P. Short,
Meadow House,
Norfolk Road, Turvey,
MK43 8DU
Tel: Turvey 461
Geese: Embden

Woodside Eggs Ltd,
Mancroft Road,
Aley Green, Slip End, Luton
Tel: Luton (0582) 841044
Ducks: Aylesbury, Khaki
Campbell, Indian Runner,
Welsh Harlequin, Silver
Appleyard
Geese: White hybrids

Berkshire
Mrs Fennie Parkinson,
The Close, Bradfield College,
Bradfield
Tel: Bradfield 744255
Geese: Brecon and Embden

Pamela Sanderson,
Hill Farm, Newtown Common,
Newbury, RG15 9DA
Tel: Newbury (0635) 40010
Ducks: Black East Indian

Buckinghamshire
R. H. Filbee,
Town Mill,
High Street, Amersham
Geese: Toulouse
Ducks: Indian Runner

C. J. S. Marler,
The Manor House,
Weston Underwood,
Olney
Tel: Bedford 711451
Geese: Chinese, (Grey and
White)

Dr Robin Thornton,
Milton Keynes, MK13 7AB
Tel: 0908 320746
Geese: Toulouse

Cambridgeshire
M. T. Norman,
48 Long Lane, Willingham
Ducks: Indian Runner

Cheshire
AMS Turkeys Ltd,
Sandlow Green Farm, Holmes
Chapel, CW4 8AS
Tel: Holmes Chapel (0477)
32329
Turkeys: BUT

Baron Turkeys,
Thatched House Farm,
Dutton,
Warrington, WA4 4JH
Tel: 09286 305
Turkeys: BUT and Roly Poly

Mark Froggatt,
Dove House Waterfowl,
Thornecliffe Wood,
Hollingworth,
via Hyde, SK14 8NJ
Tel: Mottram 63389
Ducks: Aylesbury

J. Nicholas,
Training Farm,
Barrow, Chester
Geese: Chinese

M. Rubery,
Hillside Ornamental Fowl,
Damson Lane, Mobberley,
WA16 7HY
Tel: Mobberley 3282
Ducks: Aylesbury, Buff
Orpington, Indian Runner,
Khaki Campbell, Silver
Appleyard
Geese: Brecon Buff, Embden,
Toulouse, Roman, Chinese

SCF (Turkeys) Ltd,
Heath Hey,
Heath Lane,
Childer Thornton, Cheshire,
L66 7NN
Tel: 051 339 2543
Turkeys: Wirral White, BUT

Derek Wallis,
Forest House,
Kelsall, near Tarporley,
Tel: Kelsall 51358
Geese: Chinese, Buff,
Sebastopol

Cornwall
R. D. Ackrell,
Mennawartha, St Dennis,
St Austell, PL26 8BB

Tel: St Austell 822080
Ducks: Buff Orpington

Ms V. P. A. Knight,
Penwarden Cottage,
Golberdon, Callington,
PL17 9NF
Ducks: Rouen, Aylesbury,
Khaki Campbell, Silver
Appleyard
Geese: West of England
Pilgrim, Brecon Buff

Cumbria
Denis Vernon,
Appleby Castle Conservation
Centre, Appleby, CA16 6XH
Tel: Appleby (0930) 51402
Geese: Chinese
Ducks: Indian Runner, Cayuga
Turkeys: Pied

Devon
Nigel Bell,
Sampsons Farm, Preston,
Newton Abbot, TQ12 3PP
Tel: Newton Abbot 4913
Ducks: Welsh Harlequin,
White Muscovey
Geese: Breed not specified

E. L. M. Rolle,
Pierd Corner, Stockleigh
English, Crediton,
Tel: Cheriton Fitzpaine 387
Ducks: Muscovey
Geese: Embden

Ms S. C. Squair,
Hook Farm, Hook,
Chardstock, Axminster
Tel: South Chard 20021
Ducks: Aylesbury, Khaki
Campbell, Indian Runner

Geese: Embden, Toulouse,
Chinese

Dorset
M. G. Dennis,
West Dorset Bird Farms,
Corscombe, Dorchester,
DT2 0PA
Tel: Corscombe 337
Geese: Toulouse, Chinese
Turkeys: Breed not specified

Mrs Pat Lawrence,
West Vue, Tripps Farm,
Alweston, Sherborne
Tel: Milborne Port (0963)
250 240
Ducks: Aylesbury, Khaki
Campbell (exhibition),
Rouen Claire

Miss E. B. Leakey,
The Armitage, Dottery,
Bridport, DT6 5PU
Tel: Bridport 23109
Geese: Embden

Miss C. R. Lyall,
Warmstall House, North
Chideock,
Bridport, DT6 6JW
Tel: 0297 89387
Ducks: Muscovey, Khaki
Campbell, Welsh Harlequin

Dyfed
Ms M. Duguid,
Bwlch-tre Banau,
Porth-y-Rhyd, Llanwrda
Tel: Pumpsaint (05585) 383
Geese: Brecon Buff
Ducks: Welsh Harlequin,
Magpie

H. L. Owen,
Banc Farm, Llwyn-y-Groes,
Tregaron
Tel: 0570 45 315
Geese: Brecon Buff, Toulouse

Dr C. Rouse,
Allt-y-Golau Uchaf,
Felingwm Uchaf, Carmarthen
Tel: Carmarthen 88 455
Geese: Brecon
Ducks: Welsh Harlequin

Essex
Mrs D. R. E. Arthur,
Mount Maskall, Boreham,
Chelmsford, CM3 3HW
Tel: Chelmsford 467776
Ducks: Welsh Harlequin
Geese: American Buff

D. E. Bonnet,
Longcroft, Dowsetts Lane,
Ramsden Heath, Billericay,
Tel: Basildon 710197
Geese: Chinese-Toulouse table
crosses
Ducks: Indian Runner, Khaki
Campbell, Welsh Harlequin
and other breeds

Godfreys Turkey Farms,
Broxted, Dunmow
Tel: 0279 850 293
Turkeys: BUT
Geese: Breed not specified

Kelly Turkeys Ltd,
Springate Farm, Danbury
Tel: Danbury 3581
Turkeys: Kelly Wrolstad,
Kelly Super Mini

Glamorgan
C. O. George,

The Mount, Peterson-Super-
Ely, Cardiff
Tel: 0446 760358
Geese: Brecon Buff

Gloucestershire
Tom Bartlett,
Folly Farm,
Bourton-on-the-Water,
Cheltenham
Tel: 0451 20285
Geese: Chinese, Toulouse
Rouen
Ducks: Aylesbury, Indian
Runner

M. Evans,
26A The Avenue, Ystrad
Mynach, Hengoed,
Mid-Glamorgan, CF8 8BA
Geese: Toulouse

Mr and Mrs V. J. Shinton,
Moss Rose Cottage,
Upper Soudley, Forest of
Dean, GL14 2TY
Tel: 0594 23977
Ducks: Indian Runner

Gwynedd
Ms Alison Jones,
Ty Uchair Ucha'r Ffordd,
Nebo, Caernarfon, LL54 6BW
Tel: Penygroes 881745
Ducks: Welsh Harlequin
Geese: Brecon Buff

D. L. and M. S. Wallis,
Pantgwyn Farm, Sarn Bach,
Abersoch, LL53 7ET
Tel: 075 881 2200
Geese: Sebastopol, Chinese,
Buff

Hampshire
Ms C. Coleman,
St Clairs Farmhouse,
Droxford, Southampton
Tel: 0489 878697
Geese: Chinese, Roman

Miss P. R. Greenwood,
Cold Haze Farm,
Balmer Lawn, Brockenhurst,
SO4 7TT
Tel: Lymington 23317
Geese: Brecon Buff

Miss M. R. C. Leslie,
Arford Lodge, Headley,
Bordon, GU35 8DF
Tel: Headley Down 714184
Geese: Roman

Herefordshire
G. Allen,
Hackley Farm, Bromyard
Tel: Bromyard (0885) 83685
Geese: Brecon Buff, Roman,
Embden, Toulouse
Ducks: Muscovey

P. M. Ievers,
Windsmere, Didley, Hereford,
HR2 9DA
Tel: 098121 392
Geese: Brecon Buff, Roman

Kent
Ms F. Bradley,
12 Carlton Road, Higham,
ME3 7EB
Tel: Shorne 4150
Ducks: Buff Orpington, Khaki
Campbell, Welsh Harlequin

Cottage Farm Turkeys,
Cudham,

Sevenoaks
Tel: 0959 32506
Turkeys: Small, medium and
large strains

T. B. Dafforn,
60 Dry Hill Park Road,
Tonbridge, TN10 3BX
Tel: 0732 353661
Ducks: Rouen

Sherley Davis,
Sherleys Farm, Reach Road,
St Margarets, Dover
Ducks: Aylesbury, Khaki
Campbell, Indian Runner,
Silver Appleyard,
Welsh Harlequin, Rouen
Geese: Chinese

Ms Elizabeth Dawkins,
Chickhurst Oast, Pinnock
Lane, Staplehurst, Tonbridge,
TN12 0HD
Tel: Staplehurst 892287
Ducks: Muscovey, Buff
Orpington

Kortlang and Kortlang Ltd,
The Duck Farm,
Ashford, TN25 4PD
Tel: Ashford 23431
Ducks: Aylesbury KG1 (meat),
Mortlang Tonsel strain of
Khaki Campbell (eggs)

Ms C. D. Stewart,
East Stour Farm,
Godmersham, Canterbury
Tel: Canterbury 730516
Ducks: Aylesbury, Indian
Runner

Lancashire
G. L. and E. Atkinson and
Sons, Yeomans Farm,
Briercliffe, Burnley,
BB10 3QU
Tel: 0282 22518
Turkeys: Wold White, Kelly
Wrolstadt, Leacrofter, BUT

Ms C. J. Conlon,
Little Nook Farm, Pudding Pie
Nook Lane, Goosnargh,
Preston, PR3 2JL
Tel: Preston 862002
Geese: Roman, Chinese,
Toulouse
Ducks: Welsh Harlequin

Len Cook,
4 Clement Street,
Darwen, BB3 2SB
Tel: Darwen 773953
Geese: Toulouse, Embden

Miss E. S. Hindle,
Howgill Barn, Robin Lane,
Rimington, Clitheroe,
BBF 4EF
Tel: Gisburn 579
Geese: Roman

D. Hoyle,
4 High Hudhey, Roundhill
Road, Haslingdean,
Rossendale, BB4 5BS
Ducks: Pekin, Aylesbury,
Welsh Harlequin, Indian
Runner, Buff Orpington
Geese: Brecon Buff, Chinese,
Toulouse

M. Jackson,
Oak Lynn, Jepps Lane,

Barton, Preston, PR3 5AQ
Tel: 0772 861042
Geese: Toulouse

M. H. Stephenson,
Middle Red Lumb Farm, Red
Lumb, Rochdale, OL12 7TX
Tel: Rochdale 42034
Ducks: Rouen
Geese: Brecon Buff, Pilgrim

Edward Webster,
Wash Farm, Rainford Road,
Bickerstaffe, Ormskirk,
L39 0HG
Tel: Skelmersdale 0695 24322
Turkeys: BUT, Plumpie,
Roly-Poly, Webster Bronze 1

Leicestershire
Mrs M. Abbey,
Moat House, Bramcote Road,
Loughborough
Tel: Loughborough 214154
Geese: Chinese, Toulouse
Ducks: Indian Runner

Newbold Geese,
Newbold,
Burdon, Leics.
Tel: 0455 72363
Geese: Large frame hybrids

J. H. Percival,
58 Knighton Drive,
Stoneygate, Leicester
Tel: 0533 705573
Ducks: Indian Runner

Lincolnshire
Butterball Foods Ltd,
The Hatchery, Langworth,
Tel: Lincoln 754388
Turkeys: BUT

Cherry Valley Farms Ltd,
North Kelsey Moor,
Lincoln, LN7 6HH
Tel: Caistor (0472) 851711
Ducks: Cherry Valley duckling

Middlesex
Ms S. P. Melville,
Castlebrook, Moatside,
Hanworth, Feltham,
TW13 7PG
Tel: 01 890 7222
Geese: Chinese

Norfolk
B. R. Ansell,
Honkers, 44 The Street,
Rockland St Mary, Norwich,
NR14 7AH
Tel: Surlingham 654
Geese: West of England
Pilgrim

Attleborough Poultry Farms,
Butterfly Hall, Attleborough,
NR17 1AB
Tel: Attleborough (0953)
453130
Turkeys: Attleborough Small
white, Wold Gold, Wold,
Super

J. A. Gogle,
Old Hall Farm,
Mattishall, Dereham
Tel: Dereham 850214
Geese: Brecon
Turkeys: Norfolk Black

Lawrence Mack,
Ashwicken Road, East Winch,
King's Lynn, PE32 1LJ
Tel: 0553 83 402
Turkeys: Nicholas, BUT

Norfolk Geese,
Chestnut Farm,
Pulham Market, Diss
Tel: Pulham Market (037 976)
391
Geese: Legarth heavy hybrids

E. F. Shingfield Ltd,
Hall Farm,
Hingham, Norwich
Tel: 0953 850259
Geese: Hybrids

Northamptonshire
Leacroft Turkeys,
Friars Close Farm, Barnwell,
Oundle, Peterborough
Tel: Oundle (0832) 73357
Turkeys: Leacrofter, BUT

Northumberland
Mrs Fay Dickinson,
Standwell Turkey Farm,
Harlow Hill, Horsley,
Newcastle-upon-Tyne,
NE15 0QD
Tel: Wylam (06614) 2239
Turkeys: Broadacre White

Hawkhill Turkeys,
Hawkhill, Alnwick
Tel: (0665) 830659
Turkeys: BUT

Joan Mooney,
Wandy Steads Shepherds
Cottage, Edlingham, Alnwick,
NE66 4XS
Tel: Whittingham 364
Ducks: Muscovey, Rouen

Oxfordshire
Mr and Mrs Brousson,
Fenlock House, Lower Road,

Long Harborough OX7 2LN
Tel: 0993 881875
Ducks: Buff Orpington,
Muscovey, Rouen

Ms J. Lagneau,
Locak Farm, Buscot,
SN7 8DA
Tel: 0367 52684
Geese: Toulouse

Robert Longstaff,
Orchard View, Appleton Road,
Longworth, Abingdon,
OX13 5EF
Tel: 0865 820206
Ducks: Aylesbury, Indian
Runner
Geese: Brecon Buff, Toulouse

Powys
J. M. Ashton,
Red House, Hope, Welshpool,
SY21 8JD
Tel: Welshpool 4011
Ducks: Welsh Harlequin
Geese: Brecon Buff, Chinese

David Bayliss,
The Bushes, Hope, Welspool
Tel: Welshpool 2484
Ducks: Muscovy, Indian
Runner

Ms S. Hendry,
Sychpwll, Haughton,
Llandrinio, Llanymynech
Tel: Llanymynech (0691)
831024
Geese: Brecon Buff

Shropshire
Cyril Bason (Stokesay) Ltd,
Bank House, Craven Arms,

SY7 9AN
Tel: Craven Arms (05882) 3204
Turkeys: Breed not specified

Dale Turkeys Ltd,
Caynham Hatchery,
Ludlow, SY8 4LA
Tel: Ludlow (0584) 2282
Turkeys: BUT

Highline Turkeys,
Milestone Farm, Hereford
Road,
Ludlow, SY8 4AA
Tel: Ludlow (0584) 3637
Turkeys: BUT range,
Wirral White

Ms J. A. Hodges,
2 Lythwood Hall, Bayston Hill,
Shrewsbury
Ducks: Indian Runner,
Muscovey
Geese: Brecon Buff, Chinese

Somerset
O. Dermody,
Westholme, Westfield Lane,
Draycott, Cheddar, BS27 3TP
Ducks: Aylesbury, Buff
Orpington, Rouen

Michael Hancock,
Pudleigh Mill,
Combe St Nicholas, Chard,
Tel: 04606 3663
Ducks: Indian Runner

Staffordshire
Mrs A. I. Russell,
Ladyhill Cottage, Penkridge
Bank Road, Slittingmill,
Rugeley

Tel: Rugeley 2229
Turkeys: Norfolk Black

Suffolk
John Hall,
Red House Farm,
Chediston, Halesworth
Geese: Chinese, Buff, Embden
Ducks: Indian Runner, Rouen,
Silver Appleyard

T Lay,
Waveney Wildfowl, Brook
Farm, Kirby Cane, NR35 2PJ
Tel: Bungay 2703
Ducks: Muscovey

The Priory, Waterfowl Farm,
Cyder House Ixworth, Bury St
Edmunds
Tel: 0359 31122
Ducks: Indian Runner
Geese: Chinese

Surrey
D. Allpass,
Perry Hill Farm,
Worplesdon, Guildford,
GU3 3PF
Tel: Worplesdon (0483) 232249
Geese: Roman

J. A. Crump,
Little Burfords,
Norwood Hill,
Horley, RH6 0ET
Tel: Norwood Hill (0293)
862 566
Ducks: Buff Orpington

Ms F. C. Douetil,
Busbridge Lakes, Hambledon
Road, Godalming
Tel: (048 68) 21955
Ducks: Welsh Harlequin

Geese: Chinese, Embden,
Roman

Miss P. Middleton,
Gorselands, Pirbright,
Woking, GU24 0DJ
Tel: Brookwood 2016
Ducks: Indian Runner, Welsh
Harlequin
Geese: Chinese

Sussex
Ms J. Featherstone,
Meadow View,
Marley Lane, Battle
Tel: Battle 3901
Ducks: Welsh Harlequin

Col. N. F. Gordon-Wilson,
Brookside Farm, Vines Cross,
Heathfield
Tel: Horam Road 2486
Ducks: Welsh Harlequin,
Indian Runner, Aylesbury,
Khaki Campbell
Geese: Chinese, Pilgrim
Brecon Buff

Ms F. V. Michel,
Nettlesworth Farm, Vines
Cross, Heathfield
Tel: Heathfield 2695
Geese: Chinese

Ms A. M. Mountain,
Twyford Farm, Horsted
Keynes
Tel: Chelwood Gate 313
Geese: Muscovey

Dr Jean Muddle,
Upper Mill, Plumpton
Lane, Plumpton, Lewes
Tel: 0273 890418

Geese: Embden, West of
England Crested
Ducks: Aylesbury
Turkeys: Buff

Mrs A. M. Osborne,
Shangri-La, Streat, Hassocks
Tel: Plumpton (0273) 890431
Geese: Brecon, Chinese,
Embden, Toulouse

SPR Poultry,
Barnham Station Car Park,
Barnham, Bognor Regis
Tel: Yapton (0243) 554006
Hybrid turkey poults, goslings
and ducklings

Warwickshire
Miss A. L. Clark,
Tudor House, Nuthurst,
Hockley Heath, B94 5PD
Tel: Lapworth 3254
Geese: Chinese
Ducks: Saxony, Blue Swedish

J. Billing,
White House Farm,
Fillongley, Coventry,
CV7 8DW
Tel: Fillongley 40230
Ducks: Buff Orpington

Ms S. M. and Miss L. J.
Goodall, Thompsons Farm,
Keresley End, Coventry
Tel: Keresley (020333) 2747
Ducks: Aylesbury, Indian
Runner, Rouen

W. R. Sumner,
Heathfield, 17 Birmingham
Road, Whitacre Heath,

Coleshill, Birmingham
Ducks: Indian Runner, Rouen

Mrs M. E. Wild,
Hazelwood Green House,
Preston Baghot,
Henley in Arden, Solihull
Tel: Claverdon 2963
Ducks: Indian Runner, Khaki
Campbell, Welsh Harlequin

Wiltshire
R. Wilson,
Bratch Cottage, West Hatch,
Tisbury
Tel: Tisbury 870048
Ducks: Khaki Campbell,
Muscovey

Worcestershire
R. C. Dunaway,
The Old Rectory, Hopton
Wafers, Cleobury Mortimer,
Kidderminster, DY14 0ND
Tel: Ludlow (0584) 890218
Ducks: Khaki Campbell,
Welsh Harlequin
Geese: Embden

Ms V. Roberts,
The Domestic Fowl Trust,
Honeybourne Pastures,
Honeybourne, Evesham,
WR11 5QJ
Tel: 0386 833083
Ducks: Aylesbury, Buff, Khaki
Campbell, Welsh Harlequin,
Silver Appleyard, Rouen,
Pekin, Indian Runner,
Muscovey
Geese: Brecon Buff, Chinese,
Roman, Pilgrim

C. J. Wilkinson,
Lydds Green, Ladywood,
Droitwich,
Tel: 0905 55251
Ducks: Muscovey, Silver
Appleyard
Geese: Brecon Buff

Yorkshire
F. H. Aston,
Foxwell Farm, East Layton,
Richmond,
Tel: Darlington 718579
Geese: Toulouse, Brecon Buff

G. Buxton,
108 Furlong Road, Bolton on
Dearne, Rotherham, S63 8HA
Tel: 0709 898528
Ducks: Muscovey, Rouen

Cartlidge Turkeys
(Breeders) Ltd,
Topcliffe Farm,
Tingley, Wakefield, WF3 1SQ
Tel: Morley 533179
Turkeys: Breed not specified

Brian Drake,
Warren House Farm,
Lindley Moor Road,
Huddersfield, HD3 3RS
Geese: Brecon Buff, Toulouse,
Embden

M. J. Haseler,
The Duck Farm, Skiff Lane,
Holme-on-Spalding Moor,
York, YO4 4AZ
Tel: Market Weighton (0696)
60763
Ducks: Aylesbury, Indian
Runner, Khaki Campbell,

Welsh Harlequin,
Silver Appleyard

David M. Jackson,
Keld, Northowram Green,
Halifax
Tel: Halifax 201612
Geese: Embden

J. L. Jones,
New Cottage, Winterburn,
Skipton
Tel: Airton 427
Ducks: Rouen, Indian Runner,
Aylesbury

Dennis Parkin,
Granary Farm,
Runtlings
Ossett, West Yorks, WF5 8JL
Tel: Wakefield 276541
Ducks: Khaki Campbell, Silver
Appleyard, Aylesbury x Pekin
Geese: White crosses

D. Powell,
Foxton Mill, Osmotherley,
Northallerton, DL6 3PZ
Tel: 0609 983377
Ducks: Muscovey

C. R. Sutcliffe,
Higher Moss Carr, Long
Lee, Keighley
Ducks: Aylesbury, Buff
Orpington, Rouen,
Welsh Harlequin
Geese: Brecon Buff, Toulouse,
Chinese

R. Tetley,
Browside, Thwaites Bank,
Keighley, BD21 4TG

Tel: 0535 605459
Ducks: Indian Runner

Scotland
Peter Blackwood,
Craigavon, Braal,
Halkirk, Caithness
Tel: Halkirk 238
Ducks: Khaki Campbell,
Rouen, Cayuga
Geese: Chinese, Roman

Fenton Barns (Scotland) Ltd,
Fenton Barns, North Berwick,
East Lothian
Tel: Kirleton (062 085) 202
Turkeys: BUT

D. M. Logan,
Dunedin, Knockmuir Brae,
Avoch, Ross-shire
Tel: Fortrose (0381) 20361
Geese: Embden, Toulouse
Ducks: Rouen, Crested Silver
Appleyard
Turkeys: Croullwitzer

Robin McEwan,
5 Thorntonloch,

Innerwick, Dunbar,
East Lothian, EH42 1QS
Tel: Innerwick 281
Geese: various breeds
Ducks: Indian Runner

Nicholas Europa Ltd,
9b North Vennel, Bourtreehill,
Irvine, Ayrshire, KA11 1NE
Tel: 0294 211495
Turkeys: Nicholas White

Ms C. Singleton,
Birsack Farmhouse, Kirellar,
Aberdeenshire, AB5 0TP
Tel: Kirellar 248
Ducks: Khaki Campbell,
Aylesbury
Geese: Brecon Buff, Roman

Sorrie Brothers,
Blackhall Road,
Middlefield, Inverurie,
Aberdeenshire, AB5 9JH
Tel: Inverurie (0467) 20417
Turkeys: Broad Breasted
White

Equipment suppliers
Drinkers and feeding equipment

Broiler Equipment Co Ltd,
Moorside Road, Winnall,
Winchester, Hants. SO23 7SB
Tel: Winchester (0962) 61701

EB Equipment Ltd,
Redbrook,
Barnsley, S. Yorks, S75 1HR
Tel: Barnsley (0226) 206896

George Elt Ltd,
Eltex Works, Bromyard Road,
Worcester, WR2 5DN
Tel: Worcester 422377

South and Western
(Agriculture) Ltd,
Pen Mill Trading Estate,
Yeovil, Somerset, BA21 5EA
Tel: Yeovil (0935) 74915

SPR Poultry Centre,
Barnham Station Car Park,
Barnham, Bognor Regis
Sussex
Tel: Yapton (0243) 554006

Houses

Specialist houses for ducks,
geese and turkeys are rare indeed
but these firms manufacture
small chicken houses which can
be adapted.

Free Range Systems Ltd,
Briery House,
134 Blackgate Lane,
Tarleton,
Preston, Lancs., PR4 6UU
Tel: Hesketh Bank (077 473)
3111

A. E. Jennings,
Iron Cross,
Salford Priors,
Evesham,
Worcester, WR11 5SH
Tel: Evesham (0386) 270321

R. and A. Lovett,
Crosslee Poultry Farm,
Bridge of Weir,
Renfrewshire,

Scotland, PA11 3RQ
Tel: Bridge of Weir (0505)
613075

George Mixer and Co Ltd,
Catfield,
Great Yarmouth,
Norfolk, NR29 5BA
Tel: Stalham (0692) 80355

Park Lines and Co,
501 Green Lanes,
London, N13 4BS
Tel: 01 886 0071

Southern Pullet Rearers,
Barnham Station Car Park,
Barnham, Bognor Regis,
Sussex
Tel: Yapton (0243) 554006

Sussex Joinery,
Adur Works,
Tanyard Lane,
Steyning, Sussex
Tel: Steyning 814135

Woodside Eggs Ltd,
Mancroft Road,
Aley Green, Slip End,
Luton, Beds.
Tel: Luton (0582) 841044

Appendix B

What to Buy and What it Will Cost

It is impossible to say how much it will cost to establish your garden duck, geese or turkey unit. So much will depend on individual circumstances and on how much you can improvise and use existing equipment, particularly on housing. But the following list of stock, equipment and feed will give a rough idea of the money involved. These prices were those operating in the first half of 1987, so allowances will have to be made for the rate of inflation since then.

Stock

Prices for all types of stock will vary from breed to breed or strain to strain—hens will normally cost more. Poults bought outside the breeders' busy season of July to September will be 15p to 20p a bird cheaper. Delivery will put an extra 10p on the price of each bird.

Ducks (meat)	Day-olds	£24 for 12
	10 days old (no heat)	£29.50 for 12
Egg layers	Day-old	£26 for 12
	4 weeks old	£41.50 for 12
Goslings	Day-old	£4.20 each
	7 days old	£4.30 each
Turkey poults	Day-old	£20 for 12
(high season)	5 weeks old	£40 for 12

Housing

Cost of a build-it-yourself house will depend on the materials you have on hand. Purpose-built chicken houses can be adapted. As a very rough guide to prices, a 3½ ft x 2½ ft, six-bird chicken house will cost about £100 to £120 and a 5 ft x 3 ft, twelve-bird house about £170 to £180. Remember to add on the current percentage of VAT and something for delivery.

Chicken wire: (5 ft high x 50 metres x 2 in mesh)—£50 to £55
Feeders: metal £14 to £16
          plastic £8 to £9
Galvanised feed or water troughs: about £4 each

Incubators: from £70
Shavings: £3.75 a bale
Straw: £1 to £1.25 a bale

Feed
These are prices charged by the local miller and agricultural merchant and assume you collect the feed yourself. Delivery will put at least £1 on the price. Avoid the high-street pet shops if possible as they are the most expensive places to buy feed. Pellets and crumbs will cost more than mash.

		25kg
Turkeys:	Starter crumbs	£5.80
	Ten-week growers'	£5.40
	Finisher	£5.00
Ducks and geese:	Chick starter mash	£5.00
	Growers' mash	£4.20
	Layers' mash	£4.20

Appendix C
Running Faults and Cures

Problem

All stock
Loss of condition, poor growth, birds may lose weight, feathers ruffled, possibly runny droppings—white, light brown or greyish in colour.

Possible causes and cures

Could be many things, but if only one or two birds affected, isolate from the rest in a warm, dry place with simple food—like bread and milk—and plenty of water. Give worming powders if worms suspected. Also check for freshness and availability of feed. If most of the birds are affected suspect coccidiosis or blackhead (see Chapter 6).

Droppings excessively moist, but no other symptoms.	Probably no cause for concern —possibly too much salt in the ration (scraps?) or hot weather which encourages drinking. Geese are particularly prone.
Feather pecking.	Boredom, feed shortage or skin damage to one bird that encourages others to peck. Sort out cause, treat injuries with anti-peck ointment and give birds a 'hobby' (see Chapter 6).
Swollen or distended crop area.	Object stuck in crop—crop binding; or sour crop. Can be treated (see Chapter 6). Cull if it persists.
Lameness—may be signs of swelling on the foot.	Lameness through sprain or similar damage is quite common in geese—they usually recover. All stock are susceptible to actual cuts and infection of the feet which can be cured (see Chapter 6).
Drooping wing, slipped wing.	Common in geese—caused by fright or damage on catching. Not usually detrimental and often clears up. (See Chapter 6.)

Young stock

Failure to grow: sick-looking birds in first week of life; may be some deaths.	Most likely fault is shortage of feed or water, or both. Encourage birds to eat and drink (see Chapter 2). Young birds are also susceptible to draughts and damp—pneumonia and chills likely. Keep them warm, dry and draught free.

Laying ducks

Fewer eggs than expected.	Many possible reasons: (a) Pilfering. (b) Dog eating the eggs. (c) Ducks eating own

eggs. (d) Birds put off lay by fright, etc. (e) Worms or some other health problem. (f) Sour feed or lack of feed and water. (g) Very hot or very cold weather.

A lot of soft-shelled eggs.

Normal if birds coming into lay or nearing a moult. If it persists, suspect shortage of calcium and feed limestone grit (see Chapter 5).

Mis-shapen eggs.

Not serious—may clear up. Eggs fit to eat. (See Chapter 7.)

Blood on shell.

Normal, providing only smears, particularly in early lay. If excessive and persistent, locate offender and cull.

Dirty eggs in nest.

Dirty litter coupled with bird soiling nest—change litter frequently.

Blood or meat spots in yolk or white.

Common 'problem' which normally comes and goes. No cure but eggs OK to eat. (See Chapter 7.)

Part of the bird's innards protrude from the vent.

Prolapse due to strain. Can be treated but rarely a lasting cure. Best to cull. (See Chapter 6.)

Appendix D
Daily, Weekly and Occasional Tasks

Daily
Laying ducks
(a) Collect eggs immediately after letting birds out.

(b) Check the run for any 'stray' eggs.

All stock
(a) If foxes are about—shut up birds at night and release in the mornings.
(b) If feeding wet mash or scraps, ensure that it is all eaten up and wash the trough. If dry mash, check that plenty is available.
(c) Make sure that water is available at all times.
(d) Look at the birds every day for early signs of ill health. Keep an eye on the condition of the droppings and check that the bedding is clean and dry. Look for signs of feather pecking and treat accordingly (see Chapter 6).
(e) Make sure the fence is fox-proof.

Weekly

Laying ducks
(a) Check that they are all doing their job. Good layers should have scruffy feathers. Bad ones will still look sleek and shiny (see Chapter 7).

All stock
(a) Ensure that you have plenty of feed to carry you through the following week.
(b) Watch the condition of the run in bad weather. If the run is getting very muddy look at the possibility of draining it by laying cinders or providing an alternative run (see Chapter 3).
(c) Look for evidence of mice, rats or foxes and take appropriate action if necessary (see Chapter 6).
(d) Sterilise feeders and drinkers in boiling water.

Occasional
(a) Between crops of birds tidy up the run, dig it over if necessary, and clean out the house. Discard all used litter and wash the inside of the house. Creosote the outside at least once a year.
(b) Plan for the next flock and order replacement birds well in advance of needs.

Appendix E
How Much Meat Comes off Your Bird

	Duck	Goose	Turkey
Weight before killing (live-weight)	6 lb	11 lb	8 lb
Weight after plucking and bleeding	5 lb	10 lb	7 lb
Weight after gutting but including neck, heart and liver	4.2 lb	8 lb	6.2 lb
Weight after cooking	3.7 lb	7.2 lb	5.4 lb
Yield of edible cooked meat	1.5 lb	3.2 lb	2.7 lb

(This table is given as a guide only and the yield will vary according to breed, sex and age.)

Appendix F
Suggestions for Further Reading

The following Government publications are prepared by the Ministry of Agriculture. They can be ordered through HMSO bookshops or by post from MAFF Publications Unit, Lion House, Willowburn Estate, Alnwick, Northumberland, NE66 2PF. In Scotland advisory literature is available on request from the agricultural colleges and from the Scottish Farm Building Investigation Unit.

Free publications (advisory leaflets)
L125 Duck keeping—table and egg production
L568 Disposal of chicks, turkey poults, etc.
L804 Hatching egg storage on the farm
L832 Disposal of poultry carcases
P341 Artificial incubation in small incubators
P397 Rearing turkeys for meat production
Welfare code for turkeys (No 4)

Priced publications
Ducks and Geese (RB70 1980) £5.75 (post and packing £1)
Turkey production—breeding and husbandry (RB242) £8.25 (p and p £1)
Turkey production—health (RB243) £5 (p and p 50p)

Index